DEVELOPMENTS IN SEDIMENTOLOGY 17

# SEDIMENTARY STRUCTURES OF EPHEMERAL STREAMS

FURTHER TITLES IN THIS SERIES

1. *L.M.J.U. VAN STRAATEN, Editor*
DELTAIC AND SHALLOW MARINE DEPOSITS

2. *G.C. AMSTUTZ, Editor*
SEDIMENTOLOGY AND ORE GENESIS

3. *A.H. BOUMA* and *A. BROUWER, Editors*
TURBIDITES

4. *F.G. TICKELL*
THE TECHNIQUES OF SEDIMENTARY MINERALOGY

5. *J.C. INGLE Jr.*
THE MOVEMENT OF BEACH SAND

6. *L. VAN DER PLAS Jr.*
THE IDENTIFICATION OF DETRITAL FELDSPARS

7. *S. DZULYNSKI* and *E.K. WALTON*
SEDIMENTARY FEATURES OF FLYSCH AND GREYWACKES

8. *G. LARSEN* and *G.V. CHILINGAR, Editors*
DIAGENESIS IN SEDIMENTS

9. *G.V. CHILINGAR, H.J. BISSELL* and *R.W. FAIRBRIDGE, Editors*
CARBONATE ROCKS

10. *P. McL. D. DUFF, A. HALLAM* and *E.K. WALTON*
CYCLIC SEDIMENTATION

11. *C.C. REEVES Jr.*
INTRODUCTION TO PALEOLIMNOLOGY

12. *R.G.C. BATHURST*
CARBONATE SEDIMENTS AND THEIR DIAGENESIS

13. *A.A. MANTEN*
SILURIAN REEFS OF GOTLAND

14. *K.W. GLENNIE*
DESERT SEDIMENTARY ENVIRONMENTS

15. *C.E. WEAVER* and *L.D. POLLARD*
THE CHEMISTRY OF CLAY MINERALS

16. *H. RIEKE III* and *G.V. CHILINGARIAN*
COMPACTION OF ARGILLACEOUS SEDIMENTS

DEVELOPMENTS IN SEDIMENTOLOGY 17

# SEDIMENTARY STRUCTURES OF EPHEMERAL STREAMS

BY

M. DANE PICARD

*Department of Geological and Geophysical Sciences, University of Utah, Salt Lake City, Utah (U.S.A.)*

AND

LEE R. HIGH Jr.

*Department of Geology, Oberlin College, Oberlin, Ohio (U.S.A.)*

ELSEVIER SCIENTIFIC PUBLISHING COMPANY
Amsterdam, London, New York, 1973

ELSEVIER SCIENTIFIC PUBLISHING COMPANY
335 JAN VAN GALENSTRAAT, P.O. BOX 1270, AMSTERDAM, THE NETHERLANDS

AMERICAN ELSEVIER PUBLISHING COMPANY, INC.
52 VANDERBILT AVENUE, NEW YORK, NEW YORK 10017

LIBRARY OF CONGRESS CARD NUMBER: 72-97433
ISBN 0-444-41100-3

WITH 139 ILLUSTRATIONS AND 16 TABLES



PRINTED IN THE NETHERLANDS

# PREFACE

*"All nature is but art, unknown to thee."*
(ALEXANDER POPE, 1733–34)

*"To travel hopefully is a better thing than to arrive."*
(ROBERT LOUIS STEVENSON, 1811)

Although there is a large literature on sedimentary structures in general, there are only a few papers on the sedimentology of ephemeral stream deposits. We became interested in ephemeral streams by chance. One summer morning in 1966 we became lost between Lysite and Arminto, Wyoming, and naturally ended the day by being stopped by a flood on Alkali Creek that was washing over the road. The frustration at being stopped merged into the exhilaration of rediscovery of *Dendrophycus*. From then on, we were hooked on flash floods and their deposits. As a direct result of that wet day (and serendipity), we moved erratically toward the completion of this book.

We have enjoyed the last half dozen years when the streams were flowing. There have been long and wet hours. For several moments a touch of danger has even appeared, because one of us is inept in the water. Mostly, we hope that we have advanced man's knowledge of streams by at least a centimeter.

We wish to thank the many people who gave us invaluable assistance. Parts or all of the manuscript were read at various stages by: David W. Andersen, Jonathan H. Goodwin, J.C. Harms, Earle F. McBride, Ernest P. Otto, and Charles R. Williamson. This does not make them responsible for the validity of the treatment and interpretations, and in some instances we did not follow their good advice. Typing of various drafts was done by V.R. Picard, Julie Theriot, Susan Valencia, and Jeanne Bowman. Hubert Bates, Thomas Clark, and Jay Bassin provided laboratory assistance. George Adkins, David W. Andersen, Chris Burke, Robert B. Halley, Rick Klippstein, and John Whitney were capable field assistants. To each of these individuals we express our sincere appreciation.

Photographic services were performed by Medical Illustrations and Photography, College of Medicine, University of Utah.

Financial assistance that made this study possible came from a Frederick Cottrell grant from the Research Corporation (Grant to High), the National Science Foundation (Grant GA-12570 to Picard), and the University of Utah Research Fund (Grant to Picard). In addition, Oberlin College provided support for the field assistance of Adkins and Halley, and Klippstein was aided by an Undergraduate Research Participation grant from the National Science Foundation (6Y-7405).

M. DANE PICARD
LEE R. HIGH, JR.

*To*
*Ginny and Lynn*

CONTENTS

*Chapter 1*

# INTRODUCTION

Ephemeral streams, dry washes, gullies, draws and arroyos are common names for a major element of Western landscapes. By definition ephemeral streams are characterized by short periods of flow, following local and intense rainfall, and alternating with long periods in which the channel is dry (Fig.1,2). Most ephemeral streams probably experience several flows each year, ranging in magnitude from large, destructive flash floods (Fig.3, 4) to trickles of water. Flood occurrences are erratic, but over a period of years flow frequencies are predictable. To date, most interest in ephemeral streams has centered on engineering problems and the destructive aspects of flash flooding. Our interests, however, are centered on the sedimentology of ephemeral streams.

The principal purpose of this book is to describe, illustrate and explain the sedimentary structures that are developed along ephemeral streams. The locale of the study is the Uinta Basin, northeastern Utah, where a wide variety of ephemeral streams are present (Fig.5). We also have studied similar streams in Wyoming and Colorado. Some of the specific variables of these streams are: channel size and discharge, channel shape and intrenchment, nature of load (amount of load, sediment size, sorting, composition), regional slope, and drainage basin characteristics (Fig.6–11).

However, beyond the limited goal of presenting an atlas of structures, there is one of more general significance. Through this book we hope to support and encourage the current trend of viewing sedimentary rocks as the products of ancient geologic processes rather than as interesting, isolated objects. Recent laboratory studies, many of which are cited in the following pages, have led to more precise interpretations of sedimentary structures and bedding types. The application of these studies to sedimentary rocks and ancient environments is sometimes clouded by jargon and excessive mathematics. Too frequently these reports appear to be written for hydraulic engineers rather than geologists.

Ephemeral streams form a middle ground between flume experiments and field studies of sedimentary rocks (Picard, 1970,a,b; Picard and High, 1971). Basically, ephemeral streams are natural, full-scale features that contain sedimentary deposits closely resembling sedimentary rocks. In common with flumes, flow in ephemeral streams is episodic and each event is limited in its duration. Thus,

Fig.1. Flooding on small stream flowing into Steinaker Reservoir in background.

Fig.2. Channel of small stream flowing into Steinaker Reservoir in background. Photo was taken on the day after the flood. Note high-water mark on sagebrush.

Fig.3. Flooding on Red Creek, western Uinta Basin. View of typical, large, meandering ephemeral stream during flood. Banks are retreating by undercutting and slumping. View looking downstream. Note well-developed antidunes.

Fig.4. Closeup of Fig.3. Note secondary currents moving onto bar in lower part of picture. Main current is from right to left.

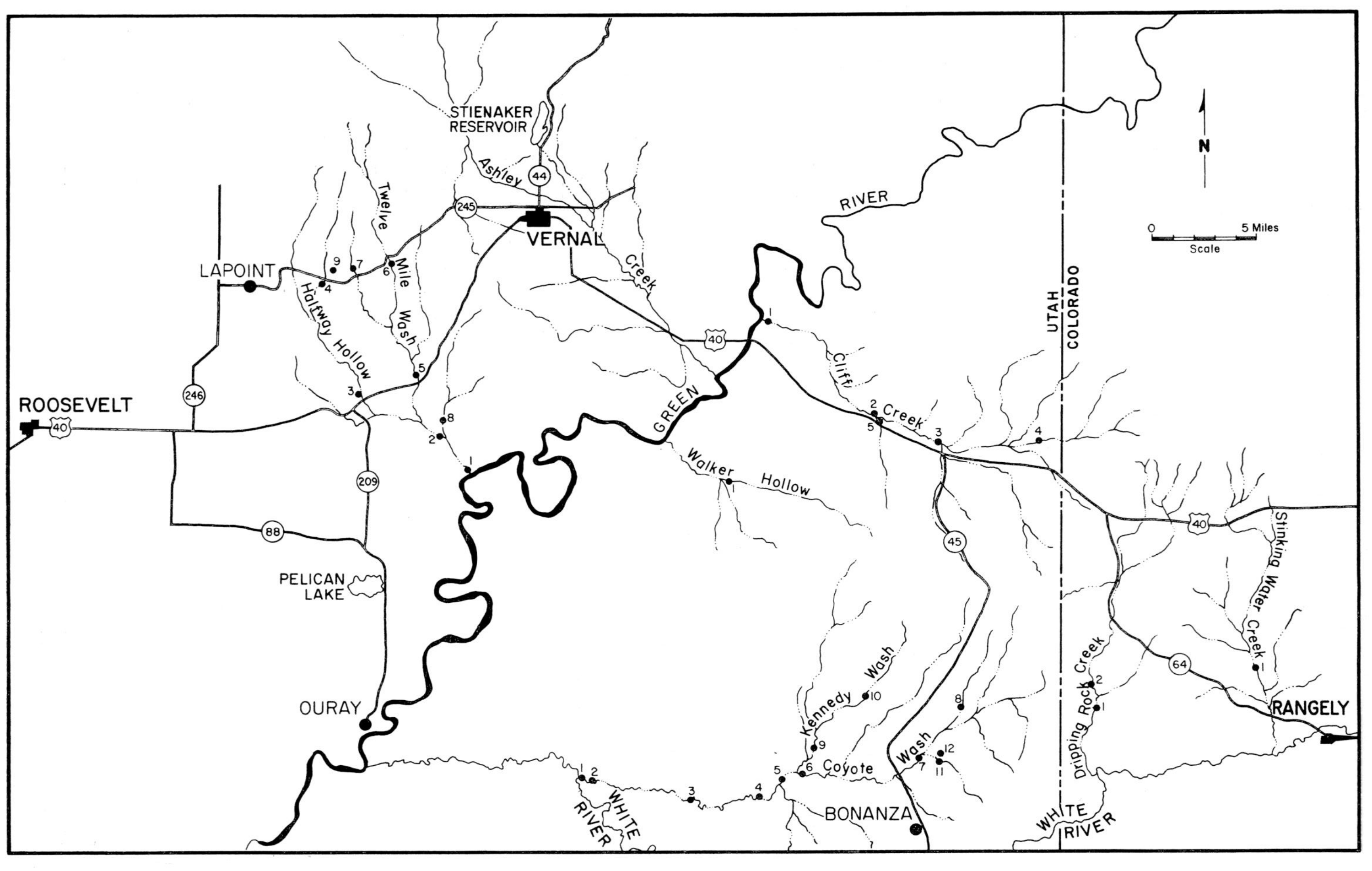


Fig.5. Index map showing locations of streams, stations along streams, and localities discussed in text.

observation of cause and effect relations are possible in ephemeral stream deposition. Furthermore, parameters such as channel size, discharge, amount of sediment load, bed roughness, velocity, grain size, slope and so forth vary from flood to flood and from stream to stream. Ephemeral streams therefore offer some of the experimental controls and repeated runs of laboratory flumes but retain natural conditions and products that can be related to sedimentary rocks.

## FORMAT AND SCOPE OF BOOK

The format of this book requires some explanation. In chapters that follow, sedimentary features are divided into categories according to a partly genetic classification (Table I). Thus, sedimentary structures are divided into erosional, transportational and post-depositional categories. Bedding types are considered separately. Although there is overlap between sedimentary structures and bedding types (as with cuspate ripples and micro cross-stratification, for example), the separation of these two features presented fewest practical difficulties. Within each chapter structures are arranged so that related forms are adjacent.

### Name and description

The first item given for each sedimentary structure is its name. The name we use is listed first in capitals and common synonyms are given within parentheses. The list of synonyms is not exhaustive and many readers will know some of the structures by other names. It is our intention to list only those synonyms that are found frequently in the literature or that have some special significance for the structure. It was not deemed necessary to include specific references to these names: most of them are in Shrock (1948), Dmitrieva et al. (1962), Pettijohn and Potter (1964), Gubler et al. (1966), and Conybeare and Crook (1968). The reader should note that the same synonym frequently appears with more than one structure. This multiple listing merely indicates either the general use of catch-all terms or imprecision and misidentification in some previous studies. With one exception, no new structures are named and we have adopted the most precise term in common use. Most names are descriptive; genetic names are avoided except where usage is strong or the feature is distinctive. Thus "rain drop impression" is retained as a term unlikely to cause difficulties while "current ripple mark" is discarded as being too general. Confusing or cumbersome terminology also is avoided. Non-informative Greek-letter names are left to their gentle and hopefully rapid expiration.

Each sedimentary structure is described concisely in terms of its general appearance and dimensions. Where significant, mean values and ranges are included. Because long descriptions quickly deaden understanding, and because

TABLE I

Classification of sedimentary structures

---

*I. Bedding-plane features*

A. Erosional
  1. current marks
    a. scour marks
      (1) current scours (scour holes, grooves, rills, incipient rib-and-furrow, micro-terraces, fluted steps, dendritic marks)
      (2) obstacle scours (scour holes, crescent scours)
    b. tool marks
  2. miscellaneous features: armored mud balls

B. Transportation and deposition
  1. ripple marks
    a. single set
      (1) curved crest (cuspate, sinuous)
      (2) straight crest (linear asymmetric, secondary, eolian, lag, longitudinal)
    b. multiple sets
      (1) simple (multiple secondary, chevron, rhomboid)
      (2) complex (cuspate and secondary, linear and secondary)
  2. lineation
    a. streaming lineation
    b. parting lineation
  3. imbrication

C. Post-depositional
  1. pre-burial
    a. desiccation features (polygonal shrinkage crack, linear-shrinkage crack, mud curl, mud pebble, reverse mud curl, earth crack)
    b. saline (salt crust, salt ridge, wrinkled surface, crystal mold)
    c. external marking (track, trail and burrow, raindrop impressions, textured surface)
    d. biologic (algal mat, algal mound)
  2. post-burial
    a. compaction (gas bubble, sand volcano, convolute bedding, flame structure, bioturbation)

*II. Internal-bedding types*

A. Planar
  1. horizontal parallel
  2. horizontal discontinuous
  3. lenticular
  4. graded

B. Cross-stratified
  1. micro cross-stratification
  2. festoon

TABLE I (continued)

3. ripple (type-A, type-B, sinusoidal)
4. inclined (scour fill, channel fill)
5. low-angle wedge
6. avalanche-front
7. backset (antidune)

C. Disturbed
1. slump features (convolute, flame structure, flow roll)
2. bioturbation (burrowed, root disturbed)
3. desiccation (mud-pebble, desiccation crack)
4. compaction
5. erosional features (stepped erosion surface)

*III. Sedimentation unit*

A. Point-bar sequence
B. Channel-bar sequence
C. Channel-fill sequence (gravel-dominated, sand-dominated)

the photographs serve this purpose better, this section is kept to a minimum.

**Occurrence and origin**

The relative abundance of each structure was recorded along several representative streams (Table II) and the pertinent line from this table is included with each structure. The convention followed is: *abundant* structures are present at all localities along stream reaches and are dominant; *common* structures are present at most localities; *rare* structures are isolated in their occurrence; and *trace* structures are found only after prolonged search. Information on relative abundance was collected only for sedimentary structures exposed on stream beds. Corresponding information on bedding types are not available although we viewed enough pits and cut banks to make some estimates. In addition to the estimates of relative abundance, oberservations concerning occurrence relative to position within and along the channel, associations with other structures, and sediment characteristics are included.

In discussing the origin of each structure we have not attempted a literature survey. Rather, we discuss the origins of the structures as we observed them along ephemeral streams. Where appropriate, information from flume studies and other depositional environments is included. Although the general origin of many structures, such as shrinkage cracks, is obvious, we tried to discuss the formation of specific types. In many instances these variations of general structures are significant for interpretations of depositional processes. For example, varieties of

Fig.6

Fig.7

Fig.8. Kennedy Wash. This stream is typical of large channels with small banks. These streams tend to be straight to sinuous, although some are now in the process of cutting off meanders that are no longer active. Longitudinal bars are well developed along the channel margin, as shown in this photo, and also within the channel. These channels are mostly cut into alluvium but some bedrock is exposed. View looking upstream.

ripple marks and lineations were found to reflect differences in current velocity.

We expect that our discussions of structure origins will draw some criticism. Our views of the formation of the structures are based on the observations given here. There should be no problems where our interpretations agree with common knowledge or experimental evidence. However, there are several instances where our interpretations are at variance with those of other workers. Probably, differences in interpretation arise because the same or similar structures are produced by

---

Fig.6. Channel of Cliff Creek. This view is typical of large, deeply entrenched ephemeral streams in northeast Utah and northwest Colorado. The channel is cut into alluvium and contains several indistinct terraces. Point bars are well developed. Flow is toward viewer.

Fig.7. Stinking Water Creek. This is another example of large, entrenched ephemeral streams, showing the development of meanders within the steep banks. The banks are retreating by undercutting and slumping and show abandoned meander scars of several ages. View looking downstream.

Fig.9. Halfway Hollow. Another view of a large, shallow channel showing development of small longitudinal bars. View looking upstream.

Fig.10. Dripping Rock Creek. This channel is moderately large and deeply entrenched. The course is straight to sinuous but changes to meandering upstream above the junction with a tributary that supplies the gravel. View looking downstream.

Fig.11. Small tributary to Halfway Hollow. Bars are poorly developed and the channel tends to be straight. Erosion is active on both sides of the channel. Small streams of this type may be largely cut into bedrock, rather than alluvium as pictured here. View looking upstream.

various combinations of conditions that are not repeated exactly from one example to the next.

Finally, we expect some readers to suggest that our interpretations are insufficiently rigorous. To this charge we plead no contest. Although we applaud the current emphasis on processes, and the application of mathematics and physics to stream studies, we are geologists who are striving for geologically useful interpretations. Hydrodynamic studies are left to those much better qualified for such endeavours than we are.

TABLE II

Abundance of sedimentary structures

| | Coyote-Kennedy Wash | | | | | | | | | | | | Halfway Hollow | | | | | | | | Cliff Creek | | | | | Dripping Rock | | Stinking Water | Walker Hollow |
|---|---|---|---|---|---|---|---|---|---|---|---|---|---|---|---|---|---|---|---|---|---|---|---|---|---|---|---|---|---|
| | *1* | *2* | *3* | *4* | *5* | *6* | *7* | *8* | *9* | *10* | *11* | *12* | *1* | *2* | *3* | *4* | *6* | *7* | *8* | *9* | *1* | *2* | *3* | *4* | *5* | *1* | *2* | | |
| Erosional structures | | | | | | | | | | | | | | | | | | | | | | | | | | | | | |
| Scour hole | c | r | r | r | r | r | c | t | r | t | – | c | r | r | t | r | r | r | t | – | c | r | r | r | – | r | c | t | c |
| Crescent scour | t | r | r | c | r | r | r | r | r | t | – | – | r | c | c | a | a | a | c | r | t | c | c | c | – | r | a | r | c |
| Tool mark | – | t | – | – | – | – | – | r | t | – | – | – | – | – | – | – | – | – | – | – | t | – | t | – | – | t | – | – | t |
| Groove | t | t | – | – | – | – | – | r | r | – | – | – | r | – | – | – | – | t | – | – | c | – | t | – | – | t | – | – | – |
| Rill | r | r | r | c | r | c | c | t | c | r | c | t | c | c | c | c | r | r | r | c | – | r | c | c | – | r | – | t | r |
| Incipient rib-and-furrow | r | r | r | r | a | r | c | r | r | r | – | r | r | r | – | – | – | – | – | – | – | – | – | – | – | – | – | – | – |
| Micro-terrace | r | r | c | c | c | r | r | r | c | c | r | c | c | r | c | c | c | c | r | c | r | r | c | c | – | c | t | c | r |
| Fluted step | r | r | t | r | r | t | r | t | c | r | – | r | c | t | r | c | r | r | – | r | – | – | r | r | – | c | t | r | t |
| Dendritic marks | – | – | – | – | – | – | r | – | t | t | – | – | r | r | t | t | – | r | r | r | r | – | – | r | – | c | t | – | – |
| Armored mud ball | – | – | t | – | t | – | – | r | t | c | – | – | t | – | – | – | – | r | – | – | – | r | t | r | – | – | – | t | – |
| Structures formed during transportation and deposition | | | | | | | | | | | | | | | | | | | | | | | | | | | | | |
| Cuspate ripple mark | a | a | a | a | a | c | c | c | a | a | – | c | a | c | r | t | – | t | – | r | t | t | r | r | – | t | t | – | – |
| Sinuous ripple mark | – | r | r | c | r | c | a | c | t | r | r | r | r | a | c | r | – | r | r | r | – | t | c | r | – | t | t | t | t |
| Linear asymmetric ripple mark | r | r | c | c | c | c | c | c | c | a | t | r | c | c | c | c | – | c | t | r | – | – | r | r | – | c | t | c | t |
| Secondary ripple mark | – | t | t | r | – | t | r | – | – | – | – | t | t | r | t | – | – | – | t | – | – | – | t | t | – | t | c | – | – |
| Eolian ripple mark | – | c | c | c | c | c | r | – | – | r | r | – | – | c | t | – | – | r | c | – | – | – | – | – | – | – | – | – | – |
| Lag ripple mark | – | – | t | t | t | r | r | – | t | t | t | – | – | r | c | t | – | r | c | c | – | – | – | r | – | – | – | – | – |
| Longitudinal ripple mark | c | c | a | a | c | c | c | a | a | t | – | r | c | c | t | c | c | c | r | – | c | t | c | r | – | r | c | t | c |
| Multiple secondary interference ripple mark | – | – | – | – | – | – | – | – | – | – | – | – | – | – | – | – | – | – | – | – | – | – | – | – | – | – | r | r | – |
| Chevron interference ripple mark | – | t | t | t | r | t | r | – | r | – | t | t | r | r | r | t | – | r | – | – | – | – | – | – | – | – | – | – | – |

| | | | | | | | | | | | | | | | | | | | | | | | | | | | | |
|---|---|---|---|---|---|---|---|---|---|---|---|---|---|---|---|---|---|---|---|---|---|---|---|---|---|---|---|---|
| Rhomboid interference ripple mark | – | t | – | t | c | t | t | – | r | – | t | – | – | r | t | – | – | – | – | – | – | – | – | – | – | – | – | – | – |
| Cuspate + secondary interference ripple mark | – | t | – | t | – | t | t | – | – | – | – | – | – | r | – | – | – | – | – | – | – | – | – | – | – | – | – | – | – |
| Linear + secondary interference ripple mark | – | – | – | t | – | t | t | – | – | – | – | – | t | t | – | – | – | – | – | – | – | – | – | – | – | – | – | – | – |
| Streaming lineation | r | r | r | c | r | c | c | c | r | c | a | r | t | c | a | a | a | a | a | a | – | – | a | a | – | – | r | – | – |
| Parting lineation | c | – | t | – | t | – | – | t | t | t | – | – | a | – | r | r | – | r | r | – | a | a | – | r | – | a | – | a | a |
| Imbrication | t | r | r | c | t | r | – | – | r | – | – | – | – | – | r | c | a | r | – | – | – | – | – | – | – | c | c | t | – |
| Post-depositional structures | | | | | | | | | | | | | | | | | | | | | | | | | | | | | |
| Polygonal-shrinkage crack | a | c | r | c | a | c | c | r | c | r | – | t | a | c | r | r | c | r | c | t | a | a | c | a | r | a | c | a | a |
| Linear-shrinkage crack | r | t | – | – | – | – | t | – | – | – | – | – | r | – | – | – | – | t | – | – | r | r | r | – | – | t | c | – | t |
| Mud curl | r | c | r | c | r | r | c | r | r | t | – | – | a | c | r | t | r | t | c | t | c | a | a | c | t | c | c | c | a |
| Mud pebble | – | t | – | – | – | t | – | – | – | – | – | – | r | – | – | – | – | – | – | – | – | r | – | t | – | – | – | t | – |
| Reverse mud curl | r | – | – | r | – | – | – | – | r | – | – | – | – | – | – | – | – | – | – | – | – | c | – | t | – | – | – | t | – |
| Earth crack | r | a | a | a | a | c | c | a | a | c | c | a | r | – | c | c | c | c | c | c | c | c | c | c | c | a | c | c | c |
| Salt crust | – | – | – | – | – | – | – | – | – | – | – | – | – | – | – | – | – | – | – | – | c | a | – | t | a | – | – | c | – |
| Salt ridge | – | – | – | – | – | – | – | – | – | – | – | – | – | – | – | – | – | – | – | – | – | a | – | – | a | – | – | c | – |
| Wrinkled surface | – | – | – | – | – | – | – | – | – | – | – | – | – | – | – | – | – | – | – | – | c | c | – | – | – | – | – | r | – |
| Crystal mold | – | – | – | – | – | – | – | – | – | – | – | – | – | – | – | – | – | – | – | – | – | t | – | – | – | – | – | – | – |
| Gas bubble | – | – | – | – | – | – | – | – | – | – | – | – | t | – | t | – | – | – | – | – | – | c | – | r | – | – | – | r | – |
| Sand volcano | t | – | – | – | – | – | – | – | – | – | – | – | t | – | – | – | – | – | – | – | – | t | – | – | – | – | – | – | – |
| Tracks, trails, burrows | a | r | t | – | r | t | r | – | r | – | r | r | r | t | r | r | c | c | t | r | c | r | r | c | r | c | r | c | r |
| Raindrop impression | c | c | t | r | r | t | – | t | c | – | – | r | c | t | c | r | r | r | t | – | a | c | r | c | r | r | r | c | t |
| Textured surface | c | r | c | c | c | c | a | c | c | r | r | c | t | t | c | c | a | c | t | r | r | r | r | a | r | – | – | – | t |
| Algal mat | – | – | – | – | – | – | – | – | – | – | – | – | t | – | – | – | – | – | – | – | – | t | – | – | – | t | – | – | – |
| Algal mound | – | c | c | c | c | c | a | – | c | r | c | r | – | – | c | r | – | t | r | t | c | c | a | c | c | c | c | c | c |

## Preservation of structures

The appearance of structures in sedimentary rocks is dependent on preservation as well as formation. Although we are unable to state probabilities or survival rates for different structures, observations of the mode of occurrence permit some speculations concerning the likelihood of preservation. The problem also can be approached differently; that is, an evaluation can be made of the sedimentary structures that have survived and are reported from fluvial rocks. Tables III and IV give the estimated abundances of structures as reported in the literature. Problems of structure identification and estimation prevent detailed interpretations.

Brief mention is also made of the occurrence of the same structure or variety in other depositional environments. With many of the structure varieties, we have

TABLE III

Relative abundance of stratification in fluvial deposits

| *Stratification* | *Ancient deposits* | | | | | | | | |
|---|---|---|---|---|---|---|---|---|---|
| | *Pocono Fm. (Miss.)* (Pelletier, 1958) | *Tuscarora Quartzite (Silurian)* (Yeakel, 1962; Smith, 1970) | *Old Red Sandstone (Devonian)* (Allen, 1962) | *Wood Bay Series (Devonian)* (Friend, 1965) | *Wamsutta Fm. (Penn.)* (Stanley, 1968) | *Carboniferous (E. Canada)* (Belt, 1968) | *Gartra Fm. (Triassic)* (McCormick and Picard, 1969) | *Wasatch Fm. (Eocene)* (Picard and High, 1970b) | *Summary of relative abundance* |
| Horizontal | c | x | c | x | c | c | c | c | c |
| Lenticular | r | | | | | | c | c | c |
| Graded | | | | x | a | | r–c | r | r–c |
| Micro cross-stratification | | x | | x | | | r | r | |
| Festoon cross-stratification | r | x | r | x | r | c | a | c–a | c |
| Ripple | | x | | x | c | c | r | r | r–c |
| Planar cross-stratification | a | x | c | x | r–c | c | c | r | c |
| Backset cross-stratification | | x | | | | | r | r | r |
| Disturbed | r | | | x | r | r | r | r–c | r |
| Massive | c | x | | | | | c | r | |
| Channel | c | x | c | x | a | c–a | a | c | c–a |
| Mud-pebble conglomerate | r–c | x | r | x | c | r–c | r–c | r | r–c |

TABLE IV

Relative abundance of sedimentary structures in fluvial deposits

| *Sedimentary structures* | *Ancient deposits* | | | | | | | | |
|---|---|---|---|---|---|---|---|---|---|
| | *Pocono Fm. (Miss.)* (Pelletier, 1958) | *Tuscarora Quartzite (Silurian) braided* (Yeakel, 1962; Smith, 1970) | *Wood Bay Series (Devonian) meandering* (Friend, 1965) | *Wamsutta Fm. (Penn.) floodplain, channel* (Stanley, 1968) | *Carboniferous (E. Canada)* (Belt, 1968) | *Gartra Fm. (Triassic) meandering* (McCormick and Picard, 1969) | *Wasatch Fm. (Eocene) braided, meandering* (Picard and High, 1970b) | *Old Red Sandstone (Devonian)* (Allen, 1962) | *Summary of relative abundance* |
| Erosional structures | | | | | | | | | |
| scour hole | r | x | x | c | r | r | r | c | r–c |
| crescent scour | | x | x | | | | | | |
| tool mark | | x | x | a | r–c | r | r | r | r–c |
| groove | | x | a | r–c | | | | | r |
| Depositional structures | | | | | | | | | |
| cuspate ripple mark | r | | x | | c | r | r | r | r–c |
| linear asymmetric ripple mark | r | x | x | r–c | r | r | r | r | r |
| interference ripple mark | r | | x | | r | r | r | | r |
| giant ripple mark | | | | | r | r | | | r |
| parting lineation | c | x | x | c | c | r | r–c | c | r–c |
| imbrication | | x | | | r | r | r | r | r |
| load cast | | | x | | r | r | | | r |
| flute cast | | x | x | a | r–c | | r | r | r–c |
| groove cast | | | x | a | r–c | | | | r |
| Post-depositional structures | | | | | | | | | |
| polygonal shrinkage crack | r | x | | a | c | r | r | | r–c |
| track, trail | r | x | | r | r | r | r | | r |
| bioturbation | | x | x | r | r | r | r | | r |
| raindrop impression | | | | r | r | | | | r |

been unable to confirm their existence in other settings, largely because of lack of precision in published descriptions. Nevertheless, considerations of occurrence and formation indicate that many of these varieties should be present in beds of other environments.

The interpretive significance of each structure is summarized briefly. Included are environmental, hydrologic and paleocurrent interpretations.

**References and illustrations**

For most structures a short list of pertinent references is included. However, we have not attempted to present a comprehensive bibliography. Instead the references listed are those that we believe are particularly important for that structure.

The photographs are the foundation of this book. Not only are they intended to illustrate each of the sedimentary structures, but we also hope that they will convey to the reader an impression of the whole depositional environment and the processes responsible for the formation of the structures. Individually, the photographs were selected to illustrate specific structures. In most instances several photographs are included to show the range of form, occurrence and associations. Together, all of the photographs demonstrate the rich variety of features that are present along ephemeral streams. All of the photographs were taken by us. With several exceptions, the photographs were taken specifically for this book in the two-year period 1969–1971.

*Chapter 2*

# EROSIONAL STRUCTURES

A large variety of erosional structures, ranging from simple scours and scratches to detailed rill patterns, are formed along ephemeral streams in the Uinta Basin. Several different types are generally present at each locality (Table II). With one exception, all of the erosional structures are depressions into the stream bed and can be classified according to the classification of sole markings proposed by Dzulynski and Walton (1965). The single exception is armored mud balls. Although similar in many respects to mud pebbles, a post-depositional structure, armored mud balls are formed by currents that gouge lumps of mud from the stream bed. Consequently, armored mud balls are included in this chapter as a dominantly erosional feature. Table V shows the erosional structures of ephemeral streams that were observed.

In general, erosional structures are present only in trace amounts; crescent scours and micro-terraces are the most abundant (rare to common), and tool marks, grooves, and incipient rib-and-furrow are the least abundant (trace amounts to not present).

TABLE V

Classification of erosional structures (after Dzulynski and Walton, 1965)

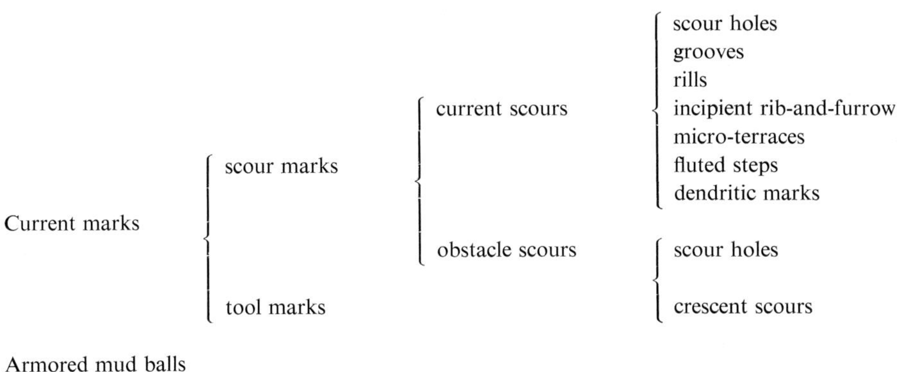

| | | | |
|---|---|---|---|
| Current marks | scour marks | current scours | scour holes |
| | | | grooves |
| | | | rills |
| | | | incipient rib-and-furrow |
| | | | micro-terraces |
| | | | fluted steps |
| | | | dendritic marks |
| | | obstacle scours | scour holes |
| | | | crescent scours |
| | tool marks | | |
| Armored mud balls | | | |

The ultimate origin of all erosional structures is stream scouring. The actual erosion is the result of several different mechanisms, including stream currents, current-generated microturbulence, obstacle-generated microturbulence, and the abrasive action of current-transported debris and sediment. Knowledge of the first two categories is largely derived from theoretical and experimental studies; and the detailed operation of these processes in geologic settings is uncertain. In contrast to these difficult topics, scour around obstacles has been the subject of numerous empirical studies relating to bridge piers (see references in Karcz, 1968). The formation of scratches by transported particles is apparent.

Most of the erosional structures reported here are bedding-plane features formed during late flood stages. Thus, most of the structures are soon exposed and destroyed. Although fluvial rocks contain abundant local erosion surfaces, detailed erosional structures are rare.

## SCOUR HOLE

(cut-and-fill, scour-and-fill, wash-out, ripple scour, toroid, obstacle mark)

### Description

Scour holes are ellipsoidal to irregular depressions within channels. In the ephemeral stream channels of the Uinta Basin most scours range from 15 to 60 cm in depth and diameters are several times greater. Generally, scours are featureless, but ripple marks or other structures are present locally within the depressions.

### Occurrence

The distribution of scours is spotty and no pattern of occurrence is discernable. Scours are present along most streams and there is no apparent association with

Scour hole

| c r r r r r c t r t – c | r r t r | r r t – | c r r r – | r c t c |
|---|---|---|---|---|

stream size, stream pattern or other hydraulic parameters. In most instances, the development of individual scours is determined by channel obstructions and is unrelated to properties of the stream.

### Origin

Scours are large-scale erosional features within the channel. Current separation and eddy formation around obstacles is the most common cause of scouring, although helical flow and semi-permanent vortices are also significant (Leighley, 1934; Matthes, 1947; Morisawa, 1968, pp.35–40).

These differences in origin lead to several distinct types of scour holes. Most scours are clearly associated with obstacles in the channel. Boulders, rock ledges, trees and driftwood are common obstructions in ephemeral channels. Separation of flow around obstructions establishes local eddies that can be strong enough to erode a depression around the base of the obstacle (Fig.12–14).

A second type of scour results from the unequal distribution of water velocity across a channel cross-section. Helical flow, a circulation cell near the banks with a horizontal axis parallel with the stream, is produced by pressure and velocity differences between the central part of the channel and the margins. If this cell erodes the channel, an elongate, trough-like scour between the channel and bank is produced (Fig.15).

The third type of scour results from semi-permanent vortices over point bars. Flow is deflected to the outsides of meanders by the abrupt change in direction, producing backwater eddies in the shallow, relatively calm water covering point bars. These eddies can scour depressions that characteristically contain a pinwheel-like ripple mark pattern (Fig.16). Backwater eddies in meanders may

Fig.12. Typical scour holes around boulders in stream bed, Twelvemile Wash. Flow is from top to bottom. Fifteen-centimeter ruler for scale.

Fig.13

Fig.14

Fig.15. Elongate scour that was formed by helical flow at the channel edge, Coyote Wash. Flow is from bottom to top.

become detached and move downstream, forming scours that are not associated with channel obstructions.

Fig.13. Large scour hole below ledge of sandstone, Dripping Rock Creek. Flow is from right to left. Note the linear ripple marks that are upstream from the ledge and on the banks of the scour hole. Secondary interference ripple marks are developed in lower left. Crescent scours are evident in the foreground. Rippled sandstone ledge is part of the fluvial Uinta Formation (Eocene). Marking pen for scale (12.5 cm) in center right on sandstone ledge.

Fig.14. Scour around tree, Cliff Creek. Free development of meander in background is prevented by tree. Flow is toward viewer.

Fig.16. Scour hole and pin-wheel ripple pattern, Halfway Hollow. Overall flow is from top to bottom with counter-clockwise flow in the scour. A point bar is visible on the lower right, and a cut bank to the top. Length of scour is about 120 cm.

## Preservation and occurrence in sedimentary rocks

Scour holes are favorably dispersed for preservation in rock sequences. Although many scours are destroyed when subsequent floods erode the channel, preservation by sediment filling is common. Such filled scours then become the cut-and-fill structures that are common in fluvial rocks. Molds of scour holes have been called toroids (Dorr and Kaufman, 1963), which superficially resemble patterned cones (Boyd and Ore, 1963) formed by upwelling springs.

It should be noted that scours, especially those produced by backwater eddies unrelated to channel obstructions, can be preserved as the lower boundaries of trough cross-stratification.

### Other occurrences

Scour holes are common features in numerous streams (Harms et al., 1963, pp.573–574); Williams and Rust, 1969, pp.656–657; McGowen and Garner, 1970, pp.82, 103). Therefore, the presence of this structure is of little significance in differentiating ephemeral from other types of streams. Furthermore, scours also are formed in marine environments, especially on tidal flats and in wave-agitated near-shore areas. Although scours are not diagnostic of depositional environment, they do record the formation of local eddies and are probably restricted to shallow water. The general scouring of the bottom by large, deep rivers during flood stages is a different process that is more related to channel formation than to individual scour holes (Leopold et al., 1964, pp.227–241).

### Significance

Scour holes are not indicative of environment but are common only in fluvial and tidal settings. Scours probably are more abundant in ephemeral and braided streams than in meandering streams. Scours may form at any position along a stream but are most abundant in upper reaches where the velocity is greatest. Most scour holes are caused by turbulence over local obstructions in the channel. The geometry of the scours has no directional significance.

### Selected references

Dorr, J. A. and Kaufman, E. G., 1963. Rippled toroids from the Napoleon Sandstone member (Mississippian) of southern Michigan. *J. Sediment. Petrol.*, 33: 751–758.

Karcz, I., 1968. Fluvial obstacle marks from the wadis of the Negev (southern Israel). *J. Sediment. Petrol.*, 38: 1000–1012.

## CRESCENT SCOUR

(current crescent, crescentic scour mark, crescent cast, horse-shoe flute cast, wash-over crescent)

### Description

Crescent scours are small, lunate depressions that are concave downstream. In size, crescent scours generally range from 1 to 15 cm in diameter and are less than

3 cm in depth. Crescent scours usually contain a small pebble or other obstruction within the curvature of the crescent and a trailing longitudinal ripple.

## Occurrence

Crescent scours are found along most streams. The structure is particularly abundant, however, along the upper reaches of streams where gravel and sand are mixed. Where the sediment is well-sorted, crescent scours are formed sporadically around the infrequent large particles.

Crescent scour

| | | | | |
|---|---|---|---|---|
| t r r c r r r r r t – – | r c c a | a a c r | t c c c – | r a r c |

Crescent scours are formed in both sand and mud. Crescents in sandy sediment tend to be larger and more irregular than those developed in mud. Crescents in mud are generally restricted to bars and floodplains.

## Origin

Crescent scours form as a result of local small-scale turbulence around obstacles. Crescents are not just small scour holes. The shape of crescent scours is regular and current oriented, in contrast to the shape of scour holes. Flow separation around small bed projections establishes turbulence on the upstream side of the obstruction. The zone of turbulence extends around the sides of the particle. The area immediately downstream of the obstacle is usually protected and a longitudinal ripple may be deposited. Because of the small size of the bed projection, turbulence strong enough to invade the lee shadow would also be strong enough to remove the obstacle. With larger obstacles, strong turbulence without transportation is possible, resulting in scour on all sides and producing a scour hole. Thus, crescent scours are always located upstream of the obstruction and the "horns" point downstream. In contrast, the location and geometry of scour holes is irregular.

Fig.17. Crescent scours in channels sands, Halfway Hollow. Flow is from left to right. Streaming lineation and lag ripples are visible in the background. The longitudinal ripples project downstream from the pebbles. (Hammer for scale.)

Fig.18. Crescent scours on rippled sand on channel bar, Halfway Hollow. Flow fans from upper left to bottom and right. Sinuous and cuspate ripples cover a bar surface that is cut by several rills in the foreground. Scale is 15 cm.

Fig.17

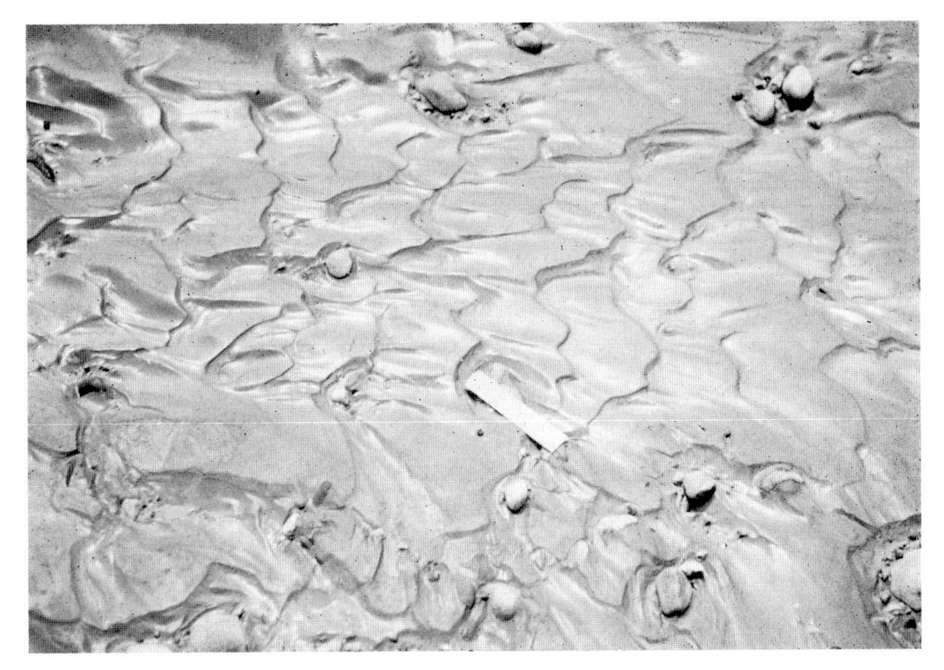

Fig.18

Fig.19. Horse-shoe crescent scours in cohesive mud on high bar, Halfway Hollow. Flow is from left to right. Note the regularity of form in comparison with larger crescent scours developed in channels. The pits in surface are produced by embedded flotsam eventually removed by the wind. Penny for scale.

In sand, crescents tend to be associated with planar bedding surfaces (Fig.17), but they are not excluded from rippled surfaces (Fig.18). Planar beds with associated streaming lineation and lag ripples indicate relatively rapid current velocities (see p.149), probably transitional into the upper-flow regime. In contrast, rippled surfaces form under lower stream velocities. Thus, the association of crescent scours with planar surfaces, upper-flow regime structures and coarse sediment suggests that crescent scours are indicative, but not diagnostic of, relatively high stream energy.

An exception to the foregoing is the occurrence of small, horse-shoe crescent scours on high bars and floodplains. These forms develop in cohesive mud and are characterized by their shape, size and regularity (Fig.19). Horse-shoe crescent scours form during flood recession just prior to exposure. Following deposition of mud, thin sheets of water draining from the bars scour horse-shoe crescents around the few obstructions. Because some of the particles are stranded flotsam (berries, nuts, woodchips, and so forth), the water sheets are shallow and the particles project above the surface. The resulting bow wave produces the horse-shoe scour. This origin probably explains the regularity of this variety of crescent scour.

## Preservation and occurrence in sedimentary rocks

As with most bedding plane structures in ephemeral channels, the preservation of most crescent scours is unlikely. Subsequent floods erode and backfill the channel, thereby destroying most of the structures on the surface. Some crescents may be buried and preserved, however, especially during channel aggradation or abandonment. The preservation of horse-shoe crescents on bars and floodplains is more likely. The cohesive mud substrate of these areas is more resistant to erosion than is the sand and gravel of the channel, enhancing the chances of survival until burial.

## Other occurrences

Crescent scours are present in all stream types, although they are probably most abundant in ephemeral streams. The structure also is formed on tidal flats and along beaches in the swash zone. Similar structures also are found in wind-blown sediment.

## Significance

The direction of current flow is toward the concave side. In sandy fluvial sediments, crescent scours are indicative of relatively high current velocities if they are associated with planar bedding. Crescent scours are most abundant at the proximal ends of streams. Horse-shoe crescents in mud are indicative of high bar or floodplain settings.

## Selected references

Allen, J. R. L., 1965. Scour marks in snow. *J. Sediment. Petrol.*, 35: 331–338.

Karcz, I., 1968. Fluviatile obstacle marks from the wadis of the Negev (southern Israel). *J. Sediment. Petrol.*, 38: 1000–1012.

Peabody, F. E., 1947. Current crescents in the Triassic Moenkopi Formation. *J. Sediment. Petrol.*, 17: 73–76.

# TOOL MARK
(prod, drag, bounce, brush, slide mark)

## Description

Tool marks are shallow, linear, discontinuous depressions that occur singly or in sets of two or three. Individual marks rarely exceed 30 cm in length, but tool marks can be repeated at close intervals downstream for great distances. Most tool marks are only a few centimeters long, about 1 to 2 cm wide, and very shallow. The mark is straight or gently curved (Fig.20). A few tool marks are short and relatively deep.

**Occurrence**

Tool marks are among the least common sedimentary structures; only a few

Tool mark

| | | | | |
|---|---|---|---|---|
| – t – – – – – r t – – – | – – – – | – – – – | t – t – – | t – – t |

scattered examples were found. Consequently, there may not be any pattern of occurrence, other than their rarity in ephemeral streams. It is probable that some examples of the more abundant grooves, here considered to be a separate structure, were formed by tools, rather than currents. Structures were designated as tool marks only where the origin was certain. All marked surfaces of uncertain origin were classified in the more general category of grooves.

**Origin**

Tool marks that we observed are formed by floating objects that drag against the bottom. Tumbleweed, tree branches and similar material probably are responsible for the majority of the marks. Linear marks are scratched in the sediment surface by trailing branches (drag mark). Where a branch jabs into the sediment, a deeper gouge is left (prod mark). Tumbling motions and irregularities in stream depth result in discontinuous marks repeated at various intervals downstream. No bounce marks were identified.

**Preservation and occurrence in sedimentary rocks**

During each flood it is probable that many tool marks form on the bottom. Considering the large amount of debris that is swept off stream banks and the floodplain, and the saltation of large pebbles, tool marks should be produced in abundance. However, the scarcity of the structure indicates that many are destroyed by the flood that produces them. Some tool marks formed during the flood crest may be buried and preserved as the flood wanes. However, because of the mobility of sandy sediment, even during the falling water stage of floods, preservation of tool marks in abundance is unlikely. More probable is the preservation of tool marks on bar tops. After deposition of much of the suspended mud, a surge across a partially submerged bar might leave tool marks in the cohesive mud. These are the types of tool marks seen in the Uinta Basin.

Fig.20. Tool marks (above pencil), polygonal shrinkage cracks and raindrop impressions, Walker Hollow. Flow is from left to right. Pencil (17 cm) for scale.

## Other occurrences

Tool marks are important structures in turbidites and slump deposits. In other geologic settings, tool marks are rare to absent.

## Significance

In fluvial rocks tool marks are suggestive of bar or flood-plain subenvironments. Tool marks yield the trend of current flow but not the direction.

## Selected reference

Dzulynski, S. and Sanders, J. E., 1962. Current marks on firm mud bottoms. *Trans. Conn. Acad. Arts Sci.*, 42: 57–96.

## GROOVE

(current mark, erosion groove, current erosional mark, longitudinal rectilinear groove)

### Description

Grooves are short, discontinuous, shallow, linear depressions that are formed as large sets of subparallel markings. Individual grooves are usually less than 30 cm in length and 3 cm in width. Some grooves become wider downstream. Grooves are typically flat-bottomed and less than 3 cm deep. Grooves superficially resemble tool marks but differ in the following respects: (*1*) tool marks occur singly or in small sets; grooves occur in swarms; (*2*) tool marks are V- or U-shaped in cross-section; grooves are flat bottomed; and (*3*) grooves tend to flare out downstream.

### Occurrence

Grooves are a minor structure along ephemeral streams. The major control on the

Groove

| | | | | |
|---|---|---|---|---|
| t t – – – – – r r – – – | r – – – | – t – – | c – t – – | t – – – |

occurrence of this structure seems to be the nature of the substrate. Grooves were observed only in cohesive mud; the structures are not developed in loose sand or in sediment that contains appreciable gravel. Grooves can be formed on bars (Fig.21) or on the bottoms and sides of channels eroded into firm sediment (Fig.22). Meander cut banks are a common site for these latter occurrences.

### Origin

Grooves are eroded into the substrate by micro-turbulence. Allen (1969) has studied the flow conditions that produce grooves. In a series of experiments, a

Fig.21. Grooves on point bar, Twelvemile Wash. Thin mud film over fine sand has been peeled away following rewetting. Note the tendency of the grooves to flare out downstream. Flow is from right to left. Cuspate ripple marks are visible in the channel at the top of the photograph. Fifteen-centimeter ruler for scale.

Fig.22. Grooves on channel bottom, Twelvemile Wash. Sediment is compact sandy mud that was exposed by erosion on the cut bank of a meander. The grooves are more irregular and slightly more sinuous than those shown in Fig. 21. They may be transitional into Allen's (1969) "longitudinal meandering grooves". Flow is from right to left. Hammer for scale.

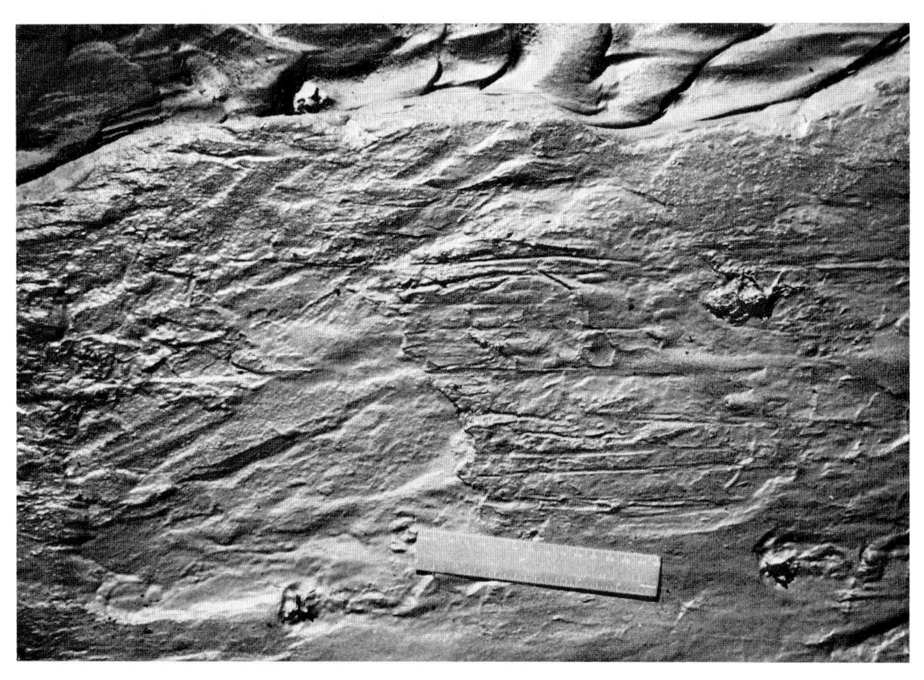

Fig.21

Fig.22

sequence of surface markings (rectilinear grooves, meandering grooves, flute marks and transverse erosional markings) was obtained as flow velocity increased. The grooves described here resemble the structures (rectilinear grooves) resulting from erosion at low flow velocities. Field evidence from ephemeral streams generally supports a low flow velocity for the formation of grooves.

The grooves shown in Fig.21 are on a high point bar. The substrate is fine sand that is covered by a thin layer of mud. The mud was deposited by a previous flood and the surface dried and hardened. A subsequent flood rewetted the surface and eroded the grooves into the mud film. Although the mud film is cohesive, it is probable that it could not have survived the vigorous scouring of high velocity flow. Thus, relatively slow flow velocity is suggested.

In contrast, the grooves shown in Fig.22 suggest higher flow velocities. At this site the substrate is older fluvial deposits, probably floodplain muds. The sediment is compact mud that is exposed on the cut bank of a meander. The sides and bottom of the channel are swept clean and no modern alluvium is present, except as point bars. The lack of sediment in the channel and the occurrence of grooves about a meter up on the cut bank suggest that current action was relatively vigorous. Although actual flow velocities are not known, it is probable that higher velocities were required to produce grooves in the semi-consolidated floodplain deposits than in the mud film covering loose sand.

### Preservation and occurrence in sedimentary rocks

Apparently grooves have not been reported from sedimentary rocks (Allen, 1969, p.613), although related erosional markings (flute marks) are common in turbidite sequences and in slump deposits. The supposed lack of grooves in rocks is difficult to believe. Grooves occur in a position at least as favorable for preservation as many other surface markings, such as rainprints, crescent scours and so forth, and they are produced in some abundance in ephemeral stream deposits. Thus, it is suggested that grooves, being minor and undistinguished bedding plane markings, are present but generally not recognized in rock sequences.

### Other occurrences

Grooves have only been described in the experimental work of Allen (1969, pp.612–613). Related structures (meandering grooves and flute marks) are found on sole markings in turbidite beds. Flute marks are particularly abundant.

### Significance

Grooves are not diagnostic of fluvial environments or subenvironments. If found in fluvial rocks, they are suggestive of relatively low flow velocities. Grooves indicate the trend but not the direction of current flow.

### Selected references

Allen, J. R. L., 1969. Erosional current marks of weakly cohesive mud beds. *J. Sediment. Petrol.*, 39: 607–623.

Dzulynski, S. and Sanders, J. E., 1962. Current marks on firm mud bottoms. *Trans. Conn. Acad. Arts Sci.*, 42: 57–96.

## RILL
(channel, 2nd- and 3rd-order channel)

### Description

Rills are small channels that dissect bars within the main channel. Generally they are less than 30 cm in width and only a few centimeters deep. The lengths of rills are dependent on bar width and range from a few centimeters to 3–4 m. Rills occur in both braided (Fig.23) and dendritic (Fig.24–25) patterns. Most are transverse to oblique in relation to the channel trend.

### Occurrence

Rills are present on the tops of most longitudinal and transverse bars within the channel. In contrast, point and lateral bars are relatively free of rills. Rill develop-

Rill

| r r r c r c c t c r c t | c c c c | r r r c | – r c c – | r – t r |
|---|---|---|---|---|

ment is variable; some bars are completely dissected while others are relatively unmodified. Rills are most common along the upper reaches of ephemeral streams.

### Origin

Rills record the erosion of small drainage channels across bars (Fig.26). Bars are constructed in the channel during flood stages. As the water level falls, the bars divide the main channel into a complex of smaller channels. An analogy may be drawn to braided rivers, although the channel-bar pattern is much simpler in ephemeral streams; generally only one bar is present at any single site along the

Fig.23

Fig.24

Fig.25. Well-developed rill-marked surface on large longitudinal bar, Cliff Creek. Rills formed by flow over bar from right to left. Flow in channel is from bottom to top. Cuspate ripple marks are shown on right. Hammer for scale.

channel. As the bars are just exposed, surges in the receding flood and turbulence wash across the bars, scouring channels. Semi-permanent drainage is incised as water drains across the bar from higher overflow channels to the lower thalweg.

### Preservation and occurrence in sedimentary rocks

Rills on exposed bars degrade with time, although surface markings may persist for several months in arid climates. Burial by younger channel or floodplain deposits

---

Fig.23. Rills on large longitudinal bar, Twelvemile Wash. Bar is between overflow channel in the background and the main channel. Flow is from top to bottom of photograph. Flow in main channel is from upper right to lower left. The rill pattern is braided. Note the crescent scours that formed around pebbles in the rills. Hammer for scale.

Fig.24. Incipient dendritic rills in longitudinal bar, Cliff Creek. Overflow channel in foreground. Flow in channel is from right to left. Flow over bar is from lower right to upper left. Rills were cut on the bar edge as surges of water drained back into the channel. Cuspate ripples were formed in the channel. Linear asymmetric ripples were developed on the bar crest in the background.

Fig.26. Small rills in the process of formation in sand bar, Cliff Creek. Flow over bar is from top to bottom. Water depth is less than 1.5 cm, yet sand grains are easily moved in rill channels, as shown by small ripples. Flow in channel is from left to right. Scale is 15 cm.

will preserve rills. In sedimentary rocks, small rill systems may be exposed on bedding planes (see dendritic marks) but exposure of most rill systems is unlikely. More probable is the preservation and exposure of discontinuous lenses of sand and gravel deposited in the rills (Williams and Rust, 1969, p.667).

**Other occurrences**

Rills are common features in braided streams; they are less common in meandering or well-regulated streams. Rills also form on tidal flats and beaches.

**Significance**

Rill marks are not indicative of fluvial environments. They are suggestive of seasonal flow variations and of proximal environments. Trends of rills are at large angles to the direction of stream flow. If two sets of rills are present, stream flow is in the direction of a line bisecting the smaller angle.

### Selected reference

Otvos, E. G., Jr., 1965. Types of rhomboid beach surface patterns. *Am. J. Sci.*, 263: 271–276.

## INCIPIENT RIB-AND-FURROW
(current break-through)

### Description

Incipient rib-and-furrow structures are compound features that result from partial erosion of rippled surfaces. The original ripple structure is partly or completely destroyed, leaving short, discontinuous ridges and hollows that display no apparent distribution pattern. Relief is moderate, generally less than 5 cm. Surfaces marked by incipient rib-and-furrow structures can extend for dozens of meters along a channel, or they may be restricted to small patches.

### Occurrence

Incipient rib-and-furrow structures are irregularly distributed along ephemeral streams in the Uinta Basin. The only apparent pattern is the association of the

Incipient rib-and-furrow

| | | | | |
|---|---|---|---|---|
| r r r r a r c r r r – r | r r – – | – – – – | – – – – – | – – – – |

structure with loose, mobile sand. However, this correlation is tenuous and may not be real. More certain is the lack of these structures in muddy or gravelly sediment.

### Origin

Although the immediate cause of this structure is the partial erosion of a rippled bed, the causes of the change from ripple formation to scour are uncertain. It is likely that the structure originates from microturbulence, possibly caused by the ripples themselves. As the flood recedes, the depth of flow decreases and the relatively large ripples become significant obstacles. In the resulting turbulence, nonselective and incomplete scour would modify the rippled surface. If the flood recedes rapidly, the scour phase might be too short-lived to result in significant ripple degradation. However, if the critical depth were maintained for a sufficient time, thorough scouring would remove all ripples. Incipient rib-and-furrow structures apparently represent a balance between these two extremes. Support for this interpretation is found at Coyote Wash (Fig.27) where the structure is present only

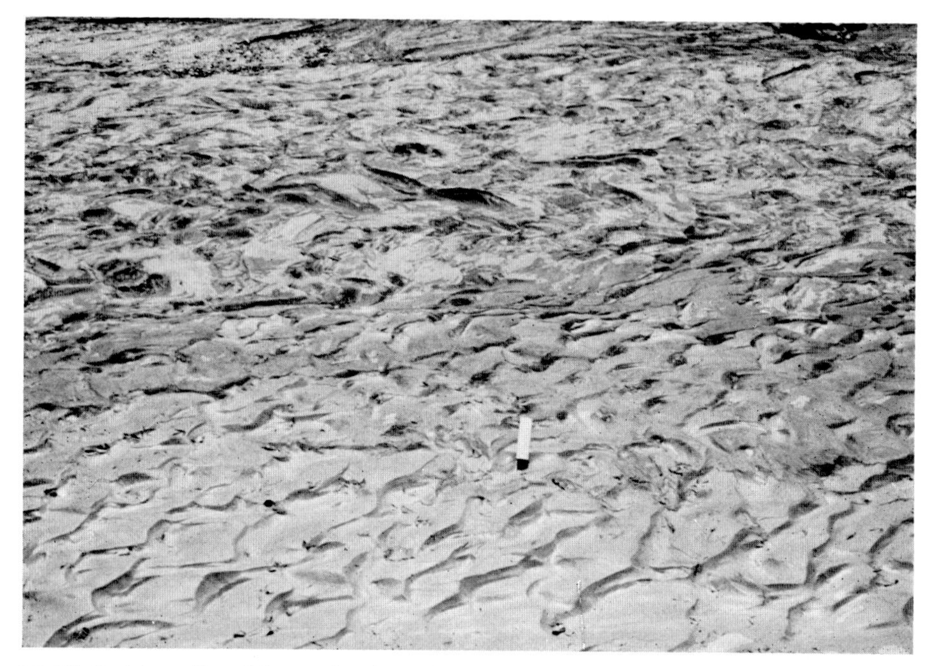

Fig.27. Incipient rib-and-furrow in channel, Coyote Wash. Flow is from left to right. Note cuspate ripples developed on longitudinal bar in foreground. Current breakthroughs are present just to right of scale. Ruler is 15 cm.

in the channel while low longitudinal bars preserve ripple-marked surfaces. Ripples were originally co-extensive across both bar and channel. However, as the water level fell, the bar was either exposed or covered to such a shallow depth as to inhibit circulation and agitation. At the same water level, the slightly deeper channel contained enough flow to scour the bottom.

**Preservation and occurrence in sedimentary rocks**

As with other large erosion surfaces, the preservation of incipient rib-and-furrow structures is possible, but would be difficult to recognize in sedimentary rocks. Current breakthroughs (erosional breaches in ripple marks) may represent the initial stage of formation of the structure. Thus, the longitudinal depression and irregularities that commonly occur with cuspate ripple-marked sandstone may result from shallow-flow microturbulence.

**Other occurrences**

Other reported occurrences of incipient rib-and-furrow structures are not known to us. Similar surfaces might be found, however, in any environment where ripple marks are formed. Rib-and-furrow is a common structure in fluvial and shallow marine deposits.

### Significance

Incipient rib-and-furrow structures are not diagnostic of environment of deposition. In fluvial rocks this structure may be suggestive of longitudinal or transverse bars. They also may define lineations which are parallel to the flow direction.

## MICRO-TERRACE
(water-level mark)

### Description

Micro-terraces are small steps cut into stream banks or bar faces. The trend is parallel with the channel and each step is horizontal. Micro-terraces are found in sets of two or more. Pairing of micro-terraces across the stream is rare. The height of each step generally is less than 5 cm and tread width ranges from 3–60 cm.

### Occurrence

Micro-terraces are widespread structures that are formed in abundance along most

Micro-terrace

| r r c c c r r r c c r c | c r c c | c c r c | r r c c – | c t c r |
|---|---|---|---|---|

ephemeral streams. Among the streams in the Uinta Basin there is no discernable pattern of distribution. Micro-terraces are preferentially developed in bar deposits probably because of the looseness of the sediment, in contrast to stream banks. Small-scale micro-terraces (Fig.28) are more common than large-scale features (Fig.29).

### Origin

Micro-terraces record the discontinuous recession of flood waters, and each level is formed during a short-lived stillstand. Micro-terraces are a small model of marine terraces on an emergent coast. Each level is an erosional structure. The tread is an abraded platform, the step a receding free face. As water level falls with the waning flood, the rate of fall is not linear. Rather, a receding flood is a series of surges, each of which is generally somewhat smaller than the preceding

Fig.28

Fig.29

surge. During surges temporary stillstand conditions exist and waves lapping the banks cut small platforms. As the surge passes, the platform is abandoned. If the next surge is shallower, a second platform and step will form below the first. Successively smaller surges produce a series of micro-terraces. An abnormally high surge obliterates earlier micro-terraces and starts the process again. As a generalization, the width of the tread is proportional to the length of stillstand.

### Preservation and occurrence in sedimentary rocks

Micro-terraces frequently are preserved beneath younger bar deposits. In pits dug across bars, micro-terraces occur as mud draped erosion surfaces with a distinctive stair-step profile. Most bars contain several of these horizons. Similar features are rare in sedimentary rocks (Ore, 1964, p.5), although some of the common erosional surfaces in fluvial units probably represent micro-terraces. The stair-step profile is diagnostic.

### Other occurrences

Micro-terraces are found on sand bars along streams of all types but are especially abundant in braided streams (Ore, 1964, p.5; Williams and Rust, 1969, p.659). Similar features are present on tidal flats (Häntzschel, 1938, p.8).

### Significance

Micro-terraces are not diagnostic of fluvial environments. In fluvial rocks, they are suggestive of bar deposits. The step is parallel with the flow direction.

### Selected references

Ore, H. T., 1964. Some criteria for recognition of braided stream deposits. *Contrib. Geol.*, 3: 1–14.

Williams, P. F. and Rust, B. R., 1969. The sedimentology of a braided river. *J. Sediment. Petrol.*, 39: 649–679.

---

Fig.28. Small-scale micro-terraces (perpendicular to pen) on lateral bar, Halfway Hollow. Channel is on right. Textured surface in left foreground; polygonal shrinkage cracks in right foreground. Length of pen is 13.5 cm.

Fig.29. Large-scale micro-terraces on longitudinal bar, Halfway Hollow. Main channel in foreground; overflow channel in background. Flow is from upper left to lower right. Length of pen is 13.5 cm.

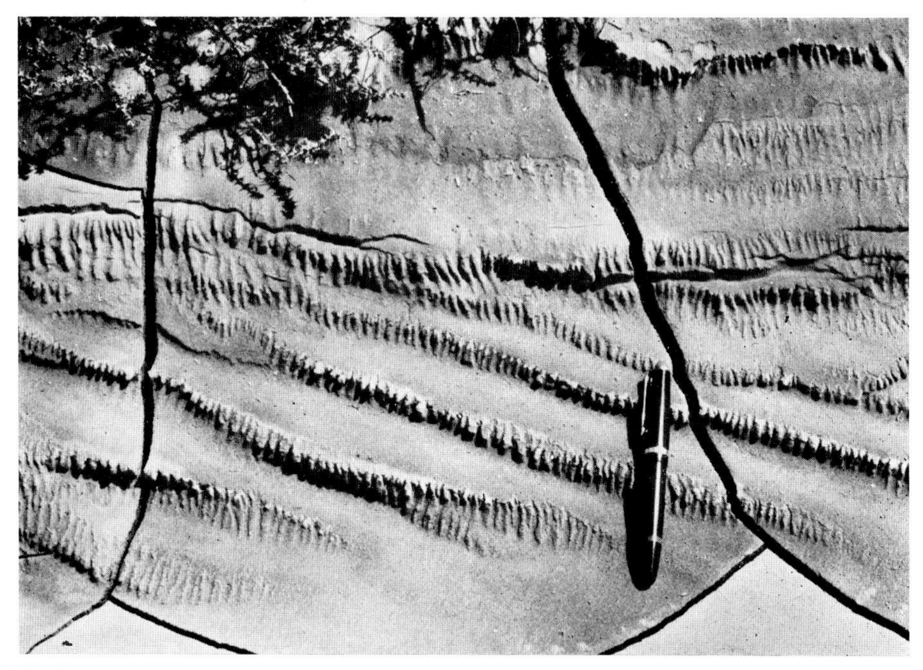

Fig.30

Fig.31

FLUTED STEPS

## Description

Fluted steps are a modified form of micro-terraces. In micro-terraces the step may be scalloped by slumping and undercutting (Fig.30) but the step is essentially unornamented. Fluted steps are characterized by vertical ridges and valleys incised into the step. Valleys are V-shaped and ridges are rounded. Spacing is fairly constant, averaging about 0.7 cm. Relief between ridges and valleys is about 0.7–1.3 cm. Generally the structure is confined to steps, although it may be poorly developed on sloping treads (Fig.30).

## Occurrence

The distribution patterns of micro-terraces and fluted steps are similar. Fluted

Fluted steps

| | | | | |
|---|---|---|---|---|
| r r t r r t r t c r – r | c t r c | r r – r | – – r r – | c t r t |

The distribution patterns of micro-terraces and fluted steps are similar. Fluted steps are as widespread as micro-terraces but are not as abundant. Fluted steps are best developed on bar sands (Fig.31), although they are also present on stream banks.

## Origin

The origin of fluted steps involves the development of ornamentation on the steps of micro-terraces. Initially, fluted steps form by the same process as micro-terraces. However, prior to the drop in water level, the step is incised with the characteristic pattern. There are two separate processes responsible for the fluted appearance. For relatively large, coarse flutes, the structure can result from small-scale slumping. As the water level falls, the free face of the step is momentarily undercut. In non-cohesive sediment, very small slumps develop all along the free face and the step

Fig.30. Fluted steps on point bar, Stinking Water Creek. The detail in these steps is atypical. Most occurrences are not so distinctly sculpted. Note the incipient dendritic drainage pattern of some of the larger flutes. Large polygonal shrinkage cracks cross the picture. Flow is from left to right. Pen for scale (13.5 cm).

Fig.31. Fluted steps on longitudinal bar, Dripping Rock Creek. This is a normal example of fluted steps. Flow in channel is from right to left. Pencil (17 cm) for scale.

is finely scalloped. For the more distinctive form of fluted step, with relatively deep V-shaped notches, the initial slumps collect water draining down the bar and the slump scars are deepened by erosion. If the amount of water is small and the sediment is cohesive, delicate flutes are formed.

### Preservation and occurrence in sedimentary rocks

The preservation of fluted steps is unlikely. The structure is delicate and rapidly degrades at surface conditions. Because the structure only develops in relatively noncohesive sediment, it is unlikely to survive until burial. Preservation is probably rare. No fluted steps have been reported previously from the fossil record. However, we recently found several fluted steps in fluvial beds of the Popo Agie Formation (Triassic) in the eastern Uinta Mountains area of northwestern Colorado. Related slump marks occur on dune deposits in the Coconino Sandstone (McKee, 1945, pp.320–323).

### Other occurrences

Fluted steps should not be restricted to ephemeral streams. Similar features are formed in the swash zone (Shrock, 1948, p.107) and are to be expected on sand bars, beaches and tidal flats.

### Significance

Fluted steps are not diagnostic of fluvial environments. If found in fluvial rocks they are suggestive of bar deposits. The trend of the step is parallel with the stream channel.

### Selected references

McKee, E. D., 1945. Small-scale structures in the Coconino Sandstone of northern Arizona. *J. Geol.*, 53: 313–325.

Shrock, R. R., 1948. *Sequence in Layered Rocks*. McGraw-Hill, New York, N.Y., 507 pp.

## DENDRITIC MARKS

(*Dendrophycus*, dendritic surge marks)

### Description

Dendritic marks are systems of branching rills cut into steep banks. Upward branching of the trunk produces a modified third- or fourth-order dendritic system. The angle of branching is generally small and each rill is straight to slightly sinuous. Rills generally cover the entire surface within the dendritic system and there are no undissected surfaces. Size of the structure varies widely, ranging from less than 3 cm (Fig.32) to 150 cm (Fig.33). The very large patterns are crude and lack the fine detail of smaller dendritic marks (Fig.34). Incomplete or incipient dendritic marks are associated with fluted steps (Fig.30).

**Occurrence**

Dendritic marks are widely distributed in ephemeral streams, although it is a minor structure in terms of abundance. The structure is formed on steep slopes in

Dendritic marks

| | | | | |
|---|---|---|---|---|
| - - - - - - r - t t - - | r r t t | - t t t | t - - t - | c t - - |

cohesive, sandy mud. The most common site is at the high water mark along restricted channel banks. Normally, dendritic marks are found in clusters upstream of channel obstructions, such as culverts or sharp bends (Fig.35). The structure is also common in tributaries just up from the junction with the main stream.

**Origin**

Dendritic marks, which were once thought to be fossils, are now known to form as surging currents erode rills on steep banks. We have observed the formation of this structure during flash floods (High and Picard, 1968). Temporary ponds are created by natural or artificial channel obstructions. Flood surges from upstream agitate the ponded water and produce considerable sloshing at the banks. Rills are eroded by the rising and falling water; if the level remains stationary for a brief period, an organized dendritic pattern forms. As the flood crest passes and the pond drains, the structure is left high on the bank. Only one level is generally produced because temporary ponding requires peak stream discharges. Normally the channels are competent to handle lower rates of flow without water backing up. The size of dendritic marks suggests the amplitude of water-level changes in the pond.

**Preservation and occurrence in sedimentary rocks**

Dendritic marks require burial to insure preservation but their occurrence high on steep banks makes preservation improbable. Dendritic marks are known from

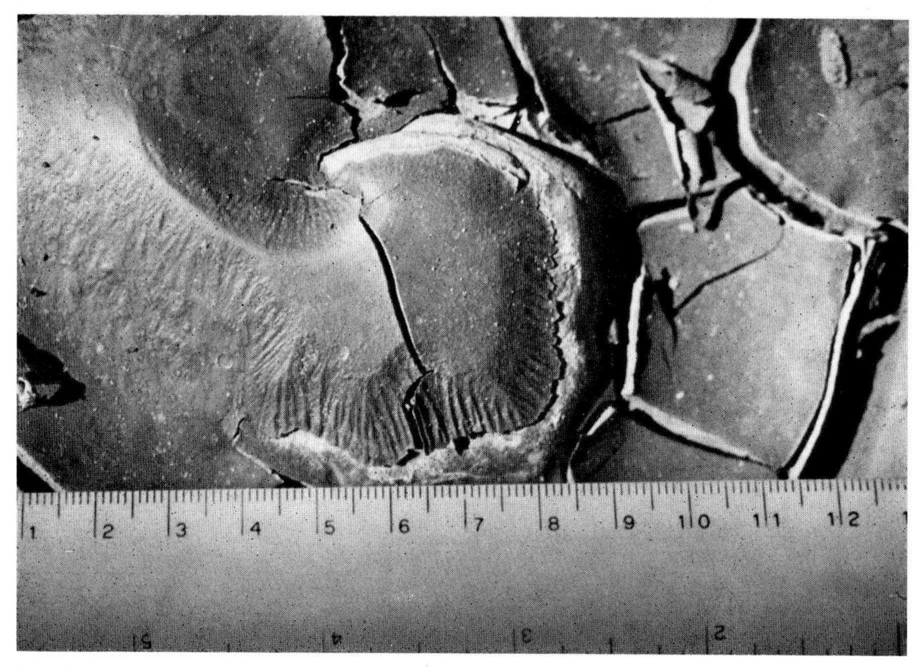


Fig.32

Fig.33

Fig.34. Normal dendritic marks on channel bank, Cliff Creek. Note the development of third- and fourth-order dendritic system. The modal grain size is silt, but there is some clay-sized material and very fine sand. Flow in channel is from left to right, but the dendritic marks were formed by surging currents moving up and down the bank face. Hammer for scale.

sedimentary rocks although the number of occurrences is small, possibly as few as six. (See High and Picard, 1968, for a review of known occurrences of this structure.) We have since found one additional occurrence in the Triassic Popo Agie Formation: High et al., 1969, pp.187–188.

### Other occurrences

Dendritic marks are found along most ephemeral streams. In addition, similar

Fig.32. Small-scale dendritic marks developed on mud coating over pebble, Little Mountain. Flow is from right to left. Crescent scour, longitudinal ridge and polygonal shrinkage cracks are also present. Scale is in centimeters.

Fig.33. Large dendritic marks on vertical channel bank, Walker Hollow. Rills are cut into mud plaster covering bottom and sides of channel. Note the large desiccation cracks and several spots where the mud layer has peeled off. Flow is from left to right. Meander and bedrock obstructions are present 16 m downstream. Flute marks are developed in lower right of photograph. Hammer for scale in right center of photograph.

Fig.35. Dendritic marks on ephemeral tributary of Alkali Creek near Arminto, Wyoming. Temporary pond was created by obstructions in culvert beneath a road. Strong surging currents are cutting the dendritic marks into the banks back of the culvert.

structures have been reported from tidal channels (Dawson, 1890, pp.614–616). Presumably a similar origin also explains these occurrences. However, High has observed a similar structure in slumped dune sand at Pentwater, Michigan. The structure therefore is not indicative of any single process of formation.

## Significance

Dendritic marks are not diagnostic of fluvial environments. If found in fluvial rocks they are suggestive of channel banks. They mark the high water levels of floods, which generally correspond to the top of the bank.

## Selected reference

High, L. R., Jr. and Picard, M. D., 1968. Dendritic surge marks *(Dendrophycus)* along modern stream banks. *Contrib. Geol.*, 7: 1–6.

Fig.36. Armored mud balls on point bar, Cliff Creek. Flow over bar is from bottom to top. Mud balls have a conglomeratic armor (in several layers) of gravel surrounding the mud. At this locality armored mud balls were abundant on the tops of bars and to a lesser extent along bar margins (toward the main channel). Pencil for scale (17 cm).

## ARMORED MUD BALLS
(clay ball, pudding ball, mud pebble, mud ball)

### Description

Armored mud balls are lumps of cohesive sediment that have a coating of small pebbles or debris on the surface (Fig.36). The mud balls are roughly spherical with diameters of a few centimeters.

### Occurrence

Armored mud balls are minor structures along the streams that we observed,

Armored mud balls

| – – t – t – – r t c – – | t – – – | – t – – | – r t r t | – – t – |
|---|---|---|---|---|

Fig.37. Oriented armored mud balls on longitudinal bar, Cliff Creek. Locality is near that shown in Fig.36. Gravel armor is generally coarser than that on mud balls in Fig.36. The distribution of the mud balls is oriented parallel with the last current to pass over the top of the bar, and the mud balls were formed (rolled-up) during this event. Note the flattened weeds that also were oriented by the last current flow (bottom to top). Hammer for scale.

occurring as isolated structures in the channel. Because of the small numbers present, no conclusions regarding distribution are possible.

**Origin**

Armored mud balls are lumps of clay or cohesive mud that have been gouged out of the stream bed or eroded from the banks by vigorous currents. The lump is rolled down the stream and becomes both rounded by abrasion and armored by accretion of small pebbles (Fig.37).

### Preservation and occurrence in sedimentary rocks

Armored mud balls are readily buried in channel deposits. If left exposed, the clay dries and the ball usually crumbles. However, if buried, the structure retains its shape. The presence of numerous armored mud balls in channel deposits exposed along stream banks indicates their likelihood of preservation. In sedimentary rocks, armored mud balls are rare, possibly because the structure is rendered unrecognizable by compaction.

### Other occurrences

Armored mud balls are abundant along some ephemeral streams, in contrast to those reported here. Mud balls that attained a diameter of 75 cm have been reported from streams in New Mexico (Leopold and Miller, 1956). Similar structures also are found along beaches, tidal channels and intertidal zones, glacial outwash areas, and in deep marine settings (Richter, 1926; Osborne, 1953; Leney and Leney, 1957; Stanley, 1964,1969).

### Significance

Armored mud balls are not diagnostic of fluvial environments. In fluvial rocks they are suggestive of channels and are indicative of strong currents.

### Selected references

Bell, H. S., 1940. Armored mud balls: Their origin, properties and role in sedimentation. *J. Geol.*, 48: 1–31.

Dickas, A. B. and W. Lunking, 1968. The origin and destruction of armored mud balls in a fresh-water lacustrine environment, Lake Superior. *J. Sediment. Petrol.*, 38: 1366–1370.

Karcz, I., 1969. Mud pebbles in a flash flood environment. *J. Sediment Petrol.*, 39: 333–337.

Stanley, D. J., 1969. Armored mud balls in an intertidal environment, Minas Basin, southwest Canada. *J. Geol.*, 77: 683–693.

*Chapter 3*

# STRUCTURES FORMED DURING TRANSPORTATION AND DEPOSITION

Many sedimentary structures in ephemeral streams are formed during the transportation and deposition of sediment. Structures of this class that we have observed include twelve varieties of ripple marks, two lineations and imbrication.

Ripple marks are difficult to classify. Early classifications equated ripple symmetry with mode of formation (symmetrical = wave, asymmetrical = current). Although the distinction between wave and current ripples is valid (Harms, 1969), confusion can result if the waves are considered to be oscillatory. In actuality, most wave ripples lack perfect symmetry (Picard and High, 1968, pp.412, 419) and result from solitary, translatory waves in shallow water. Harms' recent work (1969) on ripples is an important contribution to our understanding of their origin. Although we use a descriptive classification that is more useful in differentiating the various forms we observed, corresponding names according to Harms' genetic classification are also given (Table VI).

In contrast to biogenic or deformational sedimentary structures, features that form during transportation and deposition are potentially useful for interpreting paleohydrodynamics. Sedimentary structures frequently have been related to flow conditions in laboratory experiments (McKee, 1957, 1965; Simons and Richardson, 1961; Jopling, 1961, 1963; Simons et al., 1965; Brush, 1965; Middleton, 1967; Harms, 1969; and many others). Similar observations can also be made along modern streams (Harms and Fahnestock, 1965). Although we were unable to obtain exact measurements of flow velocities, it was possible to relate structures to flow conditions by observing modes of occurrence. Of the several variables, velocity of flow is the most significant in producing sedimentary structures; other factors such as water depth, grain size and so forth are less directly related to sedimentary structures. By observing mutual relations where more than one structure is present, the relative order of formation of structures in terms of water velocity can be determined. This order, from low to high velocity, is: parting lineation, cuspate ripples, sinuous ripples and streaming lineation. A similar sequence based on flume experiments is: plane bed, ripples, dunes and ripples, plane bed and antidunes. By comparing the structures found along ephemeral streams with those produced experimentally, we conclude that streaming lineation

TABLE VI

Classification of ripple marks

| *Geometric classification* | | | | *Genetic classification* (Harms, 1969) |
|---|---|---|---|---|
| Ripple marks | simple (single set) | curved crest | cuspate | current |
| | | | sinuous | |
| | | curved crest / straight crest | linear asymmetric | wave-dominated combined flow |
| | | | secondary | wave |
| | | straight crest | eolian | |
| | | | lag | current (?) |
| | | | longitudinal | current-dominated combined flow |
| | complex or interference (multiple sets) | single ripple type | multiple secondary | wave |
| | | | chevron | wave-dominated combined flow |
| | | | rhomboid | current-dominated combined flow |
| | | mixed ripple types | cuspate and secondary | current and wave |
| | | | linear and secondary | wave-dominated combined flow and wave |

is produced by upper-flow regime conditions, sinuous ripples are transitional between the upper- and lower-flow regimes and cuspate ripples form in the lower regime. The flow regime position of parting lineation is ambiguous.

## CUSPATE RIPPLE MARK

(current ripple mark, asymmetrical ripple mark, current mark, linguoid ripple mark, cusp ripple mark, normal ripple mark, sand wave, transverse ripple mark, crescentic ripple mark, lunate ripple mark, dune)

### Description

Cuspate ripple marks are the most common variety of current ripples (Harms, 1969). They form relatively large transverse ridges that are strongly curved in plan view.

TABLE VII

Cuspate ripple marks (82 measurements)

| *Statistics* | *Length (mm)* | *Height (mm)* | *Ripple symmetry index* | *Ripple index* |
|---|---|---|---|---|
| Mean | 147 | 22 | 3.0 | 7.6 |
| Maximum | 275 | 55 | 10.5 | 15.7 |
| Minimum | 54 | 5 | 0.9 | 2.8 |
| Standard deviation | 48 | 10 | 1.6 | 2.8 |

Where well-developed, each cusp is isolated and distinct. However, most cuspate ripples occur in short strings of three or four connected cusps (Fig.38, 39). Cuspate ripples are strongly asymmetric in profile. The average ripple symmetry index and other morphometric information are given in Table VII.

Individual cusps are concave downstream and the center of the cusp is higher than the two horns. As the cusp migrates downstream, the elevated central part leaves a ridge parallel to the direction of movement (Fig.40). The longitudinal

Fig.38. Cuspate ripple marks on channel floor, Cliff Creek. Direction of flow is toward viewer.

Fig.39

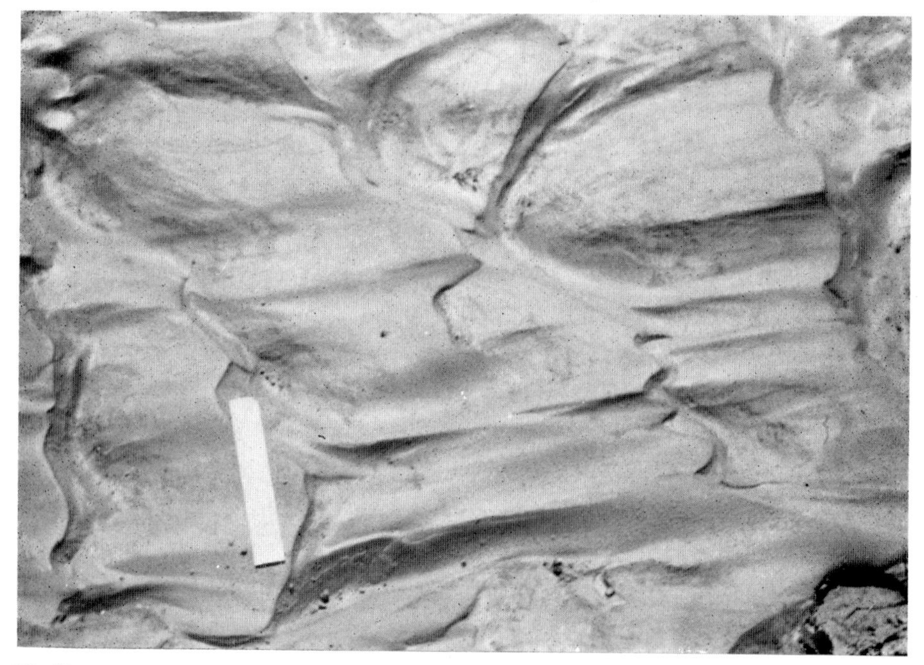

Fig.40

ridges can be so strongly developed as to dominate over the cuspate ripple mark pattern and a system of longitudinal ripples results.

Cuspate ripple marks are gradational into the higher velocity sinuous ripple marks and intermediate forms are common (Fig.40). In contrast, there does not appear to be a gradation into the lower velocity parting lineation. Cuspate ripple marks that we observed correspond to both low-energy and high-energy current ripples produced by Harms (1969, pp.365, 366), but the latter are more common.

Linguoid ripples, a special variety of cuspate ripple marks, are lobate and convex downstream. The outer rim is raised and the center depressed, yielding a tongue-like appearance. Linguoid forms occur rarely as isolated features among sets of cuspate ripples. Many linguoid ripples are associated with current break-throughs.

**Occurrence**

Cuspate ripples are among the most abundant sedimentary structures along

Cuspate ripple mark

| a a a a a c c c a a – c | a c r t | – t – r | t t r r – | t t – – |
|---|---|---|---|---|

ephemeral streams. They are formed mainly within the channel; occurrences on bars of floodplains are rare. Cuspate ripples develop in fine to coarse sand and are most abundant along the lower reaches of streams where they are associated with parting lineations and sinuous ripples.

**Origin**

Cuspate ripples are formed by moderate currents within the lower-flow regime. Water velocities near the bed are sufficient to cause saltation of most of the grains, but insufficient to support suspension.

Quantitative studies of sedimentary structures extend back to the pioneering work of H. C. Sorby. In experiments conducted in 1849, Sorby found that medium

Fig.39. Cuspate ripple marks in channel-bar complex, Ashley Creek. Ripple marks cover channel bottom and flanks of bars marginal to the channel. Bar in middleground is lower than marginal bars (foreground and background) and is covered by ripple marks. Current from right to left. Length of debris on bank (center) is about 1 m.

Fig.40. Cuspate and longitudinal ripples, Cliff Creek. Note gradation into sinuous ripples on the far left. Flow is from left to right. Fifteen centimeter ruler for scale.

to fine sand is rippled by current velocities between 15 cm/sec and 45 cm/sec (Sorby, 1908, p.180). At lesser velocities, individual grains move along the bottom without saltation, producing surface lineations (see "Parting lineation", this chapter). At greater velocities the ripples are washed away, again producing surface lineations (see "Streaming lineation", this chapter).

Similar values of current velocity were obtained by Harms (1969) in a study of ripple morphology and hydrodynamics. With fine to medium sand, current velocities less than 19 cm/sec (measured 3 cm above bed) do not result in sediment transport. Velocities between 19 and 45 cm/sec produce cuspate ripples. Velocities greater than 45 cm/sec produce dunes, defined as ripples exceeding 3 cm in height. At velocities of up to 57 cm/sec the dunes were not destroyed. The low- and high-energy varieties of cuspate ripples described by Harms (1969) are not differentiated in this study, although both types are found along ephemeral streams.

On the basis of our observations, we conclude that morphological differences in cuspate ripples partially reflect differences in current velocities. The initial form, as grains begin to saltate, is a low-relief, crenulated or lobate sand wave. Sand waves superficially resemble sinuous ripples but occur only singly, rather than in trains. Sand waves are rare and apparently form within a restricted velocity range. As velocity increases, individual lobes of sand waves are accentuated to form linguoid ripples. This phase probably corresponds to Harms' (1969) low-energy current ripples (see his plate 1 which shows lobate sand waves with linguoid ripples at top center). Finally, as flow velocity increases, full-scale microturbulence is developed and sand waves break into individual cusps that migrate independently of their former neighbors. Because this last form is by far the most abundant, it probably originates within a relative wide range of conditions.

The geometry of cuspate ripples reflects their current origin. In plan view the ripples are curved and discontinuous. This shape is so characteristic that the ripples are usually called "current ripples" (Harms, 1969). Inasmuch as the current responsible for formation can be fluvial, tidal or marine in origin and other ripple forms are also produced by currents, we prefer to use the more descriptive term "cuspate ripples". The cuspate form is the result of microturbulence. Stream flow is dominated by turbulent flow and everchanging vortices. The uniform conditions necessary to produce linearly-continuous ripple crests are generally lacking except where local conditions prevent the development of microturbulent flow. Under these conditions cuspate ripples may fail to form and sand waves can be preserved beyond their velocity field. Shallow water is the most common inhibiting factor and probably explains the preferred occurrence of sand waves on the tops of bars (see "Simple Interference Ripple Marks: Rhomboid Pattern", this chapter) and along channel margins. Flow duration may also be significant (J. C. Harms, personal communication, 1972). Bedforms require some time to come into equilibrium with flow conditions and sand waves in channels may reflect "immature" cuspate ripples preserved by the short duration of ephemeral flow.

The cross-sectional geometry of ripple marks also bears on their origin. "Current" ripples generally have low to moderate ripple indices (3–15) and high ripple symmetry indices (>3). Values for both indices for cuspate ripples from ephemeral streams in the Uinta Basin are somewhat low if compared with other "current" ripples (Tanner, 1967, p.97).

### Preservation and occurrence in sedimentary rocks

Cuspate ripple marks are infrequently preserved. Preservation is especially likely, however, if the sediment supply is great and the bottom aggrading. With lesser amounts of sediment, the ripple form may not be preserved but the internal structure may be retained as micro-cross-lamination.

### Other occurrences

Cuspate ripple marks are common structures in tidal channels and flats. The structure is also found in near-shore marine environments, especially in rip-current channels.

### Significance

Cuspate ripples are indicative of lower-flow regime currents. They are suggestive of fluvial environments, and in fluvial rocks they are indicative of channels or channel bars. The direction of current flow is normal to the ripple crest and toward the steep face. They are most abundant at the distal ends of ephemeral streams.

TABLE VIII

Sinuous ripple marks (99 measurements)

| *Statistics* | *Length (mm)* | *Height (mm)* | *Ripple symmetry index* | *Ripple index* |
|---|---|---|---|---|
| Mean | 116 | 7 | 23.6 | 23.4 |
| Maximum | 265 | 30 | 106.0 | 126.0 |
| Minimum | 45 | 1 | 1.5 | 5.0 |
| Standard deviation | 39 | 5 | 25.3 | 17.8 |

## Selected references

Allen, J. R. L., 1968. *Current Ripples–their Relation to Patterns of Water and Sediment Motion.* North Holland, Amsterdam, 433 pp.
Harms, J. C., 1969. Hydraulic significance of some sand ripples. *Bull. Geol. Soc. Am.*, 80: 363–396.
Tanner, W. F., 1967. Ripple mark indices and their uses. *Sedimentology*, 9: 89–104.

SINUOUS RIPPLE MARK
(current ripple, sand wave)

## Description

Sinuous ripple marks are a second variety of current ripples (Harms, 1969) observed along dry creeks. They are gradational into cuspate ripples (Fig.40, 41) and represent a slight change in conditions of formation. As the name implies, sinuous ripples are irregularly curved in plan view but the crest is continuous laterally for several feet (Fig.42). In addition to crest continuity, sinuous ripples (Table VIII) differ from cuspate ripples in their size and geometry. Sinuous ripples are only slightly shorter than cuspate ripples. However, the height of sinuous ripples is much less, resulting in ripple indices several times larger than those of cuspate ripples. Similarly, sinuous ripples are more asymmetrical (Fig.43).

## Occurrence

Sinuous ripples were noted only in channels; none were found on bars or floodplains. Although sinuous ripples are present along almost all streams, their

Sinuous ripple mark

| – r r c r c a c t r r r | r a c r | – r r r | – t c r – | t t t t |
|---|---|---|---|---|

distribution is not uniform. Sinuous ripples are most abundant along the middle reaches of streams. Near the mouth water velocity is not sufficient to construct this structure. Conversely, along upper reaches there may be too much gravel in the bed load for the formation of ripples. In between, where water velocity is relatively high and the load is sandy, sinuous ripples are most abundant. Sinuous ripples frequently are associated with streaming lineation; they occur less commonly with cuspate ripples.

## Origin

To our knowledge, sinuous ripples have not been produced experimentally. There-

Fig.41. Large sinuous ripples and cuspate ripples, Twelvemile Wash. Note curls of drying mud film. Ripples in sand beneath. Flow is from left to right. Hammer (about 30 cm long) for scale.

Fig.42. Small sinuous ripples, Twelvemile Wash. Note regularity of ripple crest spacing. Cuspate and linear asymmetric ripples are present at extreme right. Flow is from right to left. Hammer for scale.

Fig.43. Small sinuous ripples on tributary to Twelvemile Wash. Flow is from left to right. L = 131 mm, H = 3.2 mm, RI = 40.9, RSI = 64.8. Note the extreme asymmetry of the ripple marks and the coarse lag deposits in the troughs. Length of scale is 15 cm.

fore we are unable to assign absolute values to the water velocity. However, from observations along ephemeral streams, the position of sinuous ripples relative to other structures is clear. Sinuous ripple marks form in the lower-flow regime, but at velocities higher than cuspate ripples. Where both types of ripples are adjacent and continuous, the crests of cuspate ripples are always deflected downstream before connecting with the crests of sinuous ripples (Fig.44). Crest continuity indicates that both ripple sets are contemporaneous and related. The downstream deflection of the crests indicates a relative increase in velocity during formation of sinuous ripple marks.

Other possible controlling factors, such as grain size, depth, and flow duration can be excluded. Where sinuous and cuspate ripples occur together there is no marked difference in grain size. Thus, sinuous ripples cannot be considered to be analogues of cuspate ripples in sediment too coarse for the development of the latter. The lack of depth control is indicated by occurrences on channel floors and bar tops. Where sinuous ripples occur in one setting and cuspate ripples are adjacent in the other setting (Fig.44) cuspate ripple crests are always deflected forward onto the bar or into the channel as the case may be. Thus there is no con-

Fig.44. Gradations from cuspate ripples on bar surface (top) into sinuous ripples (middle) into streaming lineation in channel (bottom), Twelvemile Wash. Flow is from left to right. Note the deflection of ripple crests in transitions from one structure to the next. Hammer for scale.

sistent depth pattern between cuspate and sinuous, and sinuous ripples do not represent shallow-water analogues of cuspate ripples. Instead, where bars are large enough to deflect the flow, cuspate ripples will form on the bar surface and sinuous ripples form in the channel. If, on the other hand, the bars are small and are overwhelmed by the flow, velocities will be greater in the shallower waters over the bar and the pattern is reversed. These same occurrences also demonstrate that flow duration is not a factor. Both forms often occur together, excluding the possibility that sinuous ripples are nonequilibrium bedforms. Thus, velocity is shown to be the significant factor controlling the occurrence of this ripple form.

### Preservation and occurrence in sedimentary rocks

Sinuous ripples have not been reported as such from sedimentary rocks. However, some "current" ripples that have been reported are probably the sinuous variety. There is certainly no reason why sinuous ripples should not be preserved. While preservation of any bedding plane structure on the channel of an ephemeral stream is not certain, the abundance of sinuous ripples indicates that some of them will be buried and preserved.

### Other occurrences

No other occurrences of sinuous ripples are known but they probably occur with cuspate ripples in tidal and surf environments.

### Significance

Sinuous ripples are indicative of lower-flow regime conditions but flow conditions approach the transition into the upper-flow regime. They are probably good indicators of fluvial channels. Ripple crests are normal to current flow and the steep side is in the direction of flow.

### Selected references

Allen, J. R. L., 1968. *Current Ripples–their Relation to Patterns of Water and Sediment Motion.* North-Holland, Amsterdam, 433 pp.

Williams, G. E., 1971. Flood deposits of the sand-bed ephemeral streams of central Australia. *Sedimentology*, 17: 1–40.

## LINEAR ASYMMETRIC RIPPLE MARK

(wave ripple mark, parallel ripple mark, rectilinear ripple mark, combined-flow ripples)

### Description

Linear asymmetric ripple marks are the third major variety of ripple marks found along ephemeral streams. The characteristic feature of this ripple type is the great length, continuity and straightness of the crests, relative to cuspate and sinuous ripples (Fig.45). Crests are rounded or flat and bifurcations are common. The crests are oriented obliquely to the channel and the steep side is downstream and toward the bank. Morphometric information is given in Table IX.

TABLE IX

Linear asymmetric ripple marks (44 measurements)

| *Statistics* | *Length (mm)* | *Height (mm)* | *Ripple symmetry index* | *Ripple index* |
|---|---|---|---|---|
| Mean | 70 | 5 | 4.1 | 18.2 |
| Maximum | 112 | 11 | 17.0 | 65.0 |
| Minimum | 30 | 1 | 0.5 | 6.4 |
| Standard deviation | 19 | 3 | 3.4 | 13.4 |

Linear asymmetric ripple marks are a variety of wave-dominated combined flow ripples (Harms,1969).

Linear asymmetric ripple mark

| r r c c c c c c c a t r | c c c c | – c t r | – – – r – | c t c t |
|---|---|---|---|---|

## Occurrence

Linear asymmetric ripples occur widely on point, lateral, transverse, and longitudinal bars. The structure does not occur on channel floors. Linear asymmetric ripples may occur anywhere along the stream course and are not indicative of proximal-distal position.

## Origin

Linear asymmetric ripples are formed by current-generated waves that are deflected toward the banks of the channel. Waves are produced in the channel by surging flood waters. In the center of the channel these waves are too small in comparison with the turbulence of the flood to produce sedimentary structures. However, near the banks and over awash bars the waves can be the dominant type of agitation. In these relatively quiet areas the waves are refracted toward the higher topography. Because of the small size of the channels, refraction is incomplete and oblique ripples result. These ripples are combined-flow, wave-dominated ripples, according to the terminology of Harms (1969). The crest continuity and overall appearance (wave), together with the marked asymmetry (current), are indicative of the hybrid origin.

Linear asymmetric ripples can also form from refracted sinuous ripples

Fig.45

Fig.46

(Fig.46). The resulting ripples tend to be smaller (length and height) than other linear asymmetric forms and lack bifurcations. Furthermore, they are formed marginal to channels and rarely cover bar tops. Despite these differences, the general appearance and mechanics of formation are similar.

**Preservation and occurrence in sedimentary rocks**

As with other surface markings in ephemeral streams, the preservation of any single set of linear asymmetric ripples is unlikely. Scour by subsequent floods usually destroys earlier structures. However, the preferred occurrence of linear asymmetric ripples on bars slightly enhances their potential for burial and preservation. Accordingly, linear asymmetric ripples are probably more common in fluvial rocks than cuspate forms. In flood deposits linear ripples commonly are preserved as ripple stratification or climbing ripples.

**Other occurrences**

Linear asymmetric ripples are the most common type of ripple mark and they occur in many depositional environments, including deep marine, shallow marine, coastal, lacustrine and fluvial. In general appearance, these ripples are similar to eolian ripples, differing only in specific geometry.

**Significance**

Linear asymmetric ripples are not diagnostic of fluvial environments. If found in fluvial rocks, they are indicative of bars. Crests are oblique to the general direction of current flow. If a bimodal pattern is present in fluvial rocks, the paleocurrent direction bisects the acute angle. Linear asymmetric ripples are formed by current-generated waves.

**Selected references**

Harms, J. C., 1969. Hydraulic significance of some sand ripples. *Bull. Geol. Soc. Am.*, 80: 363–396.

---

Fig.45. Linear asymmetric ripples on point bar, Twelvemile Wash. Flow is from left to right. Current-generated waves are refracted toward bank in background. Note bifurcations of crests. Hammer for scale.

Fig.46. Linear asymmetric ripples on point bar, Coyote Wash. Linear ripples are gradational into sinuous ripples in center. The current was refracted onto the bar at the bottom of picture. Flow is from left to right. Fifteen-centimeter ruler for scale.

Newton, R. S., 1968. Internal structure of wave-formed ripple marks in the nearshore zone. *Sedimentology*, 11: 275–292.
Picard, M. D., 1967a. Paleocurrents and shoreline orientations in Green River Formation (Eocene), Raven Ridge and Red Wash areas, northeastern Uinta Basin, Utah. *Bull. Am. Assoc. Petrol. Geologists*, 51: 383–392.
Picard, M. D., 1967b. Paleocurrents and shoreline orientations in Green River Formation (Eocene), Raven Ridge and Red Wash areas, northeastern Uinta basin, Utah: Reply to discussion by R. A. Davis, Jr. *Bull. Am. Assoc. Petrol. Geologists*, 51: 2471–2475.
Picard, M. D. and High L. R., Jr., 1968. Shallow-marine currents on the Early (?) Triassic Wyoming shelf. *J. Sediment. Petrol.*, 38: 411–423.
Reineck, H.-E., 1961. Sedimentbewegungen an Kleinrippeln im Watt: *Senckenbergiana Lethaea*, 42: 51–67.
Reineck, H.-E., und Wunderlich, F., 1968. Zur Unterscheidung von asymmetrischen Oszillationsrippeln und Strömungsrippeln. *Senckenbergiana Lethaea*, 47: 321–345.

## SECONDARY RIPPLE MARK

### Description

Secondary ripple marks are small-scale varieties of linear asymmetric ripples (Fig.47, 48). Both patterns have long, relatively straight, continuous crests and

TABLE X

Secondary ripple marks (8 measurements)

| *Statistics* | *Length (mm)* | *Height (mm)* | *Ripple symmetry index* | *Ripple index* |
|---|---|---|---|---|
| Mean | 19 | 2 | 3.1 | 9.9 |
| Maximum | 22 | 3 | 4.5 | 13.0 |
| Minimum | 13 | 1 | 1.1 | 5.0 |
| Standard deviation | 4 | 1 | 1.4 | 2.4 |

Fig.47. Secondary ripples on lateral bar, Ashley Creek. Channel is to the left. The dark debris in the right is just below the high water limit. Flow in channel is toward viewer. Ripples are asymmetric toward the right. Quarter for scale.

Fig.48. Secondary ripples on lateral bar, Ashley Creek. Flow in channel is from left to right; direction of flow of refracted waves from bottom to top. Ripples are asymmetric toward the top. Quarter for scale.

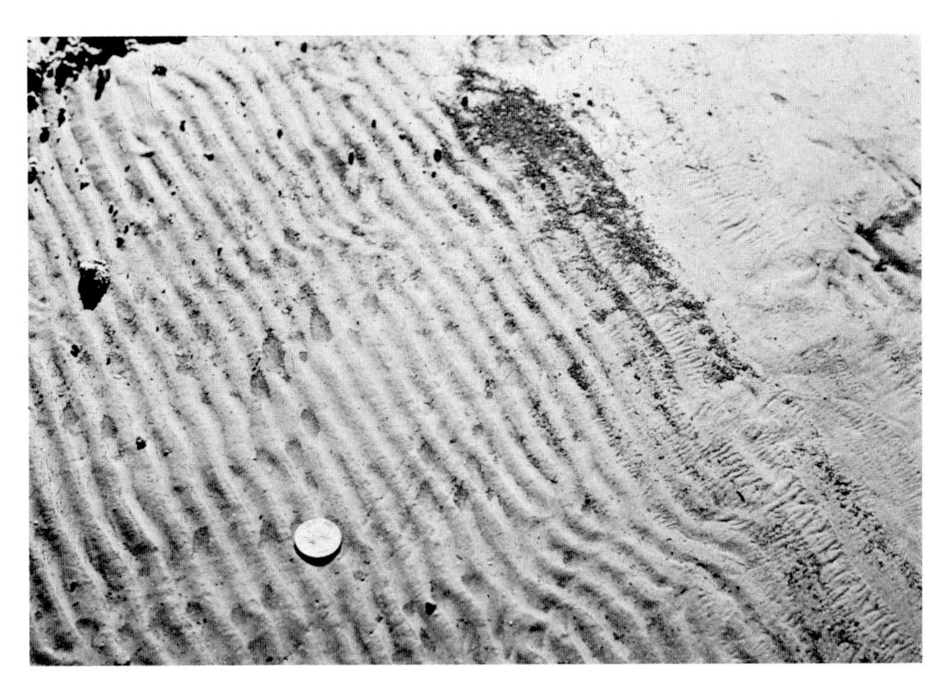

Fig.47

Fig.48

Fig.49. Refracted waves and resulting secondary ripples on bar crest. View toward center of stream. Channel flow from right to left. Refracted current from top to bottom. Note cuspate ripples on flanks of bar.

common bifurcations. Secondary ripples (Table X) differ from linear asymmetric forms in size and occurrence. The former are several times smaller and can have any orientation. Their steep faces may be onshore, offshore, upstream, downstream, or any conceivable oblique direction. Linear asymmetric ripples are often oblique (between onshore and downstream). Secondary ripples are a variety of wave ripples (Harms, 1969).

## Occurrence

Secondary ripple marks are found on channel margins and bars. Commonly, secondary ripples are associated with other type of ripples and form an interference pattern (Picard and High, 1970a). Secondary ripples form in fine sand or silt and may occur anywhere along the stream course.

Secondary ripple mark

| – t t r – t r – – – – – t | t r t – | – – t – | – – t t – | t c – – |
|---|---|---|---|---|

## Origin

Secondary ripples are developed by a variety of waves that lap up on bars and stream banks (Fig.49). These waves originate in the channels from current flow. Unlike the waves that form linear asymmetric ripples, the waves that produce secondary ripples are generally completely refracted. In addition to refraction, which produces waves moving directly toward the banks, diffraction and reflection also are significant. Diffraction of waves around the ends of bars produces oblique patterns and reflection generally produces off-bank waves. Combinations of different wave sets result in complex patterns.

## Preservation and occurrence in sedimentary rocks

Small, wave-formed ripple marks similar to secondary ripples are known in sedimentary rocks. The likelihood of preservation of this structure in fluvial sediments is similar to that of linear asymmetric ripples. The occurrence of secondary ripples on bars enhances the possibility of their burial and preservation.

## Other occurrences

Secondary ripples should be formed in any environment where primary currents are subject to refraction, diffraction or reflection. In addition to bars along meandering and braided streams, the structure probably is formed in tidal and nearshore environments.

## Significance

Secondary ripples are not diagnostic of fluvial environments. If found in fluvial rocks they are indicative of bars or channel margins. They have no directional significance, and are formed by modified, current-derived waves.

### Selected references

Newton, R. S., 1968. Internal structure of wave-formed ripple marks in the nearshore zone. *Sedimentology*, 11: 275–292.

Picard, M. D. and High, L. R., Jr., 1970a. Interference ripple marks formed by ephemeral streams. *J. Sediment. Petrol.*, 40: 708–711.

## EOLIAN RIPPLE MARK

### Description

Eolian ripples are low amplitude sand waves with straight to curved, linear, continuous, rounded crests. Bifurcations are common. In overall appearance, eolian ripples resemble linear asymmetric and secondary ripples. Eolian ripples are distinguished from other types by their mode of occurrence and high ripple index (Tanner, 1964, 1966, pp.133, 134).

### Occurrence

Eolian ripples are a minor structure along ephemeral streams. They are most abundant on the floodplains of larger streams (Fig.50). Eolian ripples are formed in loose, fine, well-sorted sand. In the Uinta Basin the floodplain alluvium along

Eolian ripple mark

| | | | | |
|---|---|---|---|---|
| – c c c c c r – – r r – | – c t – | – r c – | – – – – – | – – – – |

most streams is slightly muddy. When this poorly sorted sediment dries, a hard crust forms and prevents the construction of eolian ripples. Less commonly, the structure is found on channel bottoms as winds transport the sediment left behind by the previous flood.

### Origin

Persistent winds in deserts or semi-arid regions are common. Thus, the formation of eolian ripples requires only a suitable sediment supply. Both sediment size and sorting are critical limiting parameters. Well-sorted, loose, fine to medium sand is the optimum sediment in the Uinta Basin area. Coarser sediment is not moved by normal winds; finer sediment is carried away in suspension. The addition of even small amounts of mud binds the sediment so that upon drying wind erosion is

Fig.50. Eolian ripples on floodplain, Twelvemile Wash. Direction of transport is away from viewer.

negligible. Folk (1971, p.23) suggested that wind ripples generally are formed in well-sorted sand ranging in mean size from medium to very coarse. Eolian ripple marks with wave lengths from 0.2 to 2 cm have been found in sandy very fine gravel (Sharp, 1963) and in coarse gravel (Smith, 1965).

## Preservation and occurrence in sedimentary rocks

Eolian ripples are particularly susceptible to destruction, even more so than other structures along ephemeral streams. The reason is that these ripples develop only in loose sand. If the surface is subsequently flooded, a prerequisite to burial, even slight currents in the water will remove the ripples. Preservation is likely only if the sand supply is so large as to progressively fill a local depression with sheets of wind-blown sand.

## Other occurrences

Eolian ripples are widespread in dune and beach environments and are more typical of these settings than of fluvial deposition.

Fig.51

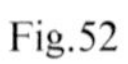

Fig.52

Fig.53. Close-up of lag ripples in figure 52. Flow is from right to left. Scale is 15 cm.

## Significance

Eolian ripples are not diagnostic of fluvial environments. If present in fluvial rocks they are indicative of ephemeral flow. Eolian ripples indicate wind direction at the time of sediment transport, and may reflect local sources of sediment rather than prevailing wind directions.

## Selected references

Bagnold, R. A., 1941. *The Physics of Blown Sand and Desert Dunes.* Methuen, London, 265 pp.
Sharp, R. P., 1963. Wind ripples. *J. Geol.*, 71: 617–636.

---

Fig.51. Lag ripples in channel, Twelvemile Wash. Flow is toward viewer. Streaming lineation is present beneath the lag ripples and is especially well-developed at the lower right. Note the cuspate ripples in depression on the left where the current fanned out and velocity decreased markedly. Pen (about 15 cm) for scale.

Fig.52. Lag ripples in channel, Twelvemile Wash. Note the coarse size of the sediment in the ripples, especially above the 15-cm scale. Lineation is well-developed on the pavement beneath the ripples. Flow is from right to left.

## LAG RIPPLE MARK
(new structure)

### Description

Lag ripples are faint stripes of coarse sediment oriented normal to the direction of current flow. The stripes are short, generally only a few inches in length, and are evenly spaced in large sets. The rippled surface can extend for tens of feet downstream. The coarse sediment that comprises lag ripples rests on top of a smooth surface of poorly sorted, muddy, gravelly sand (Fig.51–53).

Although they are formed by currents, lag ripple marks are not represented in Harms' (1969) classification.

### Occurrence

Lag ripples are always associated with streaming lineation, which is a common

Lag ripple mark

| | | | | |
|---|---|---|---|---|
| – – t t t r r – t t t – | – r c t | – r c c | – – – r – | – – – – |

structure along the upper reaches of streams. Lag ripples are developed only in the channel. The ripples are composed of well-sorted, granule-size sediment.

### Origin

Lag ripples and streaming lineation record the hydrodynamic separation of different sized material under conditions of relatively high velocity. Sand-sized sediment is transported in the upper-flow regime to produce plane beds and streaming lineation. In contrast, coarser sediment lags behind and moves by saltation in the lower-flow regime. These processes lead to two separate structures, indicative of two modes of transport and two "energy" levels, produced simultaneously by a single current.

### Preservation and occurrence in sedimentary rocks

Inasmuch as this is a new structure, lag ripples have not been reported from sedimentary rocks. However, there are no compelling reasons why the structure should not be preserved. Since planar bedding is abundant in many fluvial deposits, a close inspection of bedding surfaces should reveal the presence of lag ripples.

### Other occurrences

We are not aware of similar structures having been reported elsewhere. However, the structure should be common along most ephemeral streams and possibly beaches, where streaming lineation was originally described (Conybeare and Crook, 1968, p.25).

### Significance

It is not known if lag ripples are diagnostic of fluvial environments. In fluvial sediment they are indicative of channel conditions. Lag ripples are formed by lower-flow regime currents but they are indicative of upper-flow regime transport of the bulk of the sediment. Orientation of the structure is normal to the direction of current flow.

## LONGITUDINAL RIPPLE MARK
(sand shadow, window ridge)

### Description

Longitudinal ripple marks are short, linear, straight sand ridges that are oriented parallel to the channel. Size ranges from less than 3 cm in length to more than 30 cm. The crests of longitudinal ripples are rounded to subangular and the ripples have a high degree of symmetry. Bifurcations are absent. Longitudinal ripples can occur singly or in sets. Longitudinal ripple marks probably are a variety of current-dominated combined flow ripples (Harms, 1969).

### Occurrence

Longitudinal ripples are formed in two common settings, along the channels and bars of ephemeral streams. Sets of small longitudinal ripples are associated with

Longitudinal ripple mark

| c c a a c c c a a t – r | c c t c | c c r – | c t c r | r c t c |
|---|---|---|---|---|

cuspate and sinuous ripples (Fig.39–43), but individual longitudinal ripples occur behind channel obstacles and are associated with scour holes and crescents (Fig.54). Both types are widely distributed along ephemeral streams.

## Origin

Longitudinal ripples form by two distinct mechanisms. Those associated with channel obstacles, scours and crescents are simply lee deposits. Sediment accumulates in the current shadow downstream of an obstacle. The resulting ridge is a compound ridge with opposed foresets (Fig.54). These longitudinal ripples are most abundant along upper reaches of the stream.

Longitudinal ripples also form during the migration of cuspate and sinuous ripples. As a cuspate ripple migrates downstream, the apex of the cusp, which is the highest point along the crest, leaves a ridge behind (Fig.40). Subsequent deposition about the ridge results in an asymmetric ripple. Sets of these ripples are most abundant along the lower reaches of a stream.

## Preservation and occurrence in sedimentary rocks

Both varieties of longitudinal ripples are present in sedimentary rocks. As with other structures on bars and within the channel, preservation of any single structure

Fig.54. Longitudinal ripple on downstream side of boulder, Twelvemile Wash. Note current crescent and smaller longitudinal ripples to right and left of boulder. The internal structure of the longitudinal ripple shows opposed foresets of compound ripple. Flow is toward viewer. Length of pen is 13 cm.

is unlikely but possible. Given the large numbers of longitudinal ripples that form along ephemeral streams, some should survive.

### Other occurrences

Longitudinal ripples similar to those described here are found in sediments of other fluvial environments and in tidal-flat and channel sediments. Large longitudinal ripples formed directly by currents are present in the Wadden Sea (Van Straaten, 1951).

### Significance

Longitudinal ripples are not diagnostic of fluvial environments or subenvironments. They are indicative of current action but not the relative velocity. Crests are parallel to the flow direction.

### Selected reference

Van Straaten, L. M. J. U., 1951. Longitudinal ripple marks in mud and sand. *J. Sediment. Petrol.*, 21: 47–54.

## SIMPLE INTERFERENCE RIPPLE MARK: MULTIPLE SECONDARY, CHEVRON, RHOMBOID RIPPLE PATTERNS

### Description

Simple interference ripple marks are composed of several superimposed sets of a single ripple variety. Although all ripple types can form interference patterns, only secondary and linear asymmetric ripples were found in simple interference sets along the ephemeral streams.

Secondary, simple interference patterns are constructed of multiple sets of secondary ripples. Up to four contemporaneous sets were observed at one locality. Aside from the presence of multiple, superimposed sets, the ripples are identical with other secondary ripples.

Linear asymmetric ripples form two distinct interference patterns. The chevron pattern is formed by linear ripples that are gently curved and convex downstream (Fig.55, 56). The rhomboid pattern is formed by two overlapping sets of linear ripples migrating in different directions (Fig.57).

Multiple secondary, chevron, and rhomboid interference ripple patterns are varieties of wave, wave dominated combined flow, and current dominated combined flow ripples, respectively (Harms, 1969).

Fig.55

Fig.56

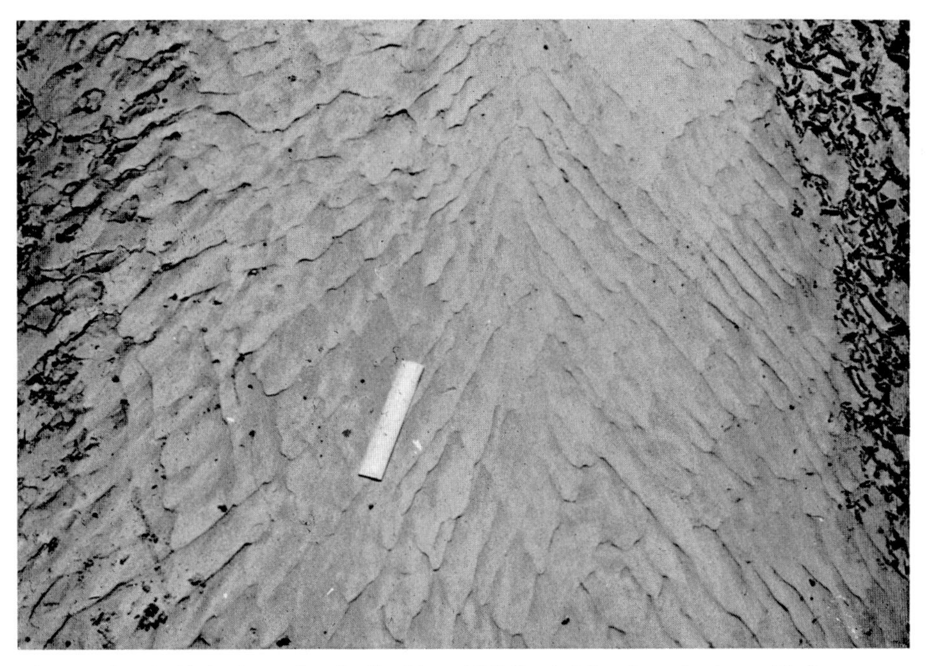

Fig.57. Rhomboid ripples on longitudinal bar, Cliff Creek. Flow in main channel is from top to bottom. The flow separates around the bar and waves wash over the bar from top right and top left. Unmodified linear asymmetric ripple pattern is visible on the right. Mud curls are marginal to the bar. Length of scale is 15 cm.

## Occurrence

Multiple secondary interference ripple patterns are rare. They are found with

Multiple secondary ripple mark

| | | | | |
|---|---|---|---|---|
| – – – – – – – – – – – – – | – – – – | – – – – | – – – – – – | – r r – |

Chevron ripple mark

| | | | | |
|---|---|---|---|---|
| – t t t r t r – r – t t | r r r t | – r – – | – – – – – – | – – – – |

---

Fig.55. Chevron ripples in overflow channel on point bar, Halfway Hollow. Ripples are of the linear asymmetric variety and are modified by flow in the narrow channel. Flow is from left to right. Hammer for scale.

Fig.56. Chevron ripples in main channel of Dripping Rock Creek. Current from left to right. Quarter for scale.

Rhomboid ripple mark

| – t – t c t t – r – t – | – r t – | – – – – | – – – – – – | – – – – |
|---|---|---|---|---|

secondary ripples along channel margins. Chevron patterns occur in small chutes either within the channel or draining large bars. Rhomboid ripples are developed on the tops of transverse and longitudinal bars within the channel.

### Origin

In the quiet water along the edge of a channel, current-generated waves are refracted, diffracted and reflected. Although the refracted waves that move bankward are generally dominant, any of the other patterns can also produce ripple sets. When this occurs, multiple interference sets of secondary ripples are formed.

Similarly, the origin of chevron ripples is straightforward. Water surges move through overflow channels on bars. Linear asymmetric ripples are formed by the wave dominated currents in these small channels. Because the channels are narrow the waves cannot move freely. Rather, the waves are retarded at each end by the shallow water. The resulting ripples mirror the distorted wave pattern and are concave upstream (Fig.55, 56). The center of the ripple moves faster than either end and, therefore, migrates farther. The result is the distinctive chevron pattern.

Rhomboid ripples are formed by the interference of two wave sets. As the current separates around a mid-channel bar, flood surges send sheets of water over the bar from two directions. Rapidly repeated surges construct a set of linear asymmetric ripple marks. Where the two sets cross, an interference pattern in the form of rhombs results (Fig.57).

### Preservation and occurrence in sedimentary rocks

Chevron ripples have not been reported in ancient rocks. Multiple secondary and rhomboid patterns occur in the fossil record but only rarely. All three of these structures are equally likely to be preserved. However, in view of the small probability for preservation of most structures in ephemeral streams and the relative scarcity of these ripple marks, they probably are minor sedimentary structures.

### Other occurrences

Rhomboid markings similar to rhomboid ripples are common to abundant along the swash zones of beaches. Multiple secondary interference ripple patterns are common in small ponds and near the banks of quiet lakes. Chevron patterns could form in tidal environments, although none have been reported.

## Significance

These ripple marks are not diagnostic of fluvial environments. If present in fluvial rocks, they are indicative of channel margins or bars and are suggestive of surging currents. Secondary sets have no directional significance. In the chevron pattern the "V" points downstream; in rhomboid ripples the "V" points upstream.

## Selected references

Hoyt, J. H. and Henry, V. J., Jr., 1963. Rhomboid ripple mark, indicator of current direction and movement. *J. Sediment. Petrol.*, 33: 604–608.

Woodford, A. O., 1935. Rhomboid ripple mark. *Am. J. Sci.*, 5(29): 518–525.

Fig.58. Complex interference ripple marks in center-right, Cliff Creek. Dominant ripples are linear asymmetric and flow is from right to left. However, note well-developed secondary ripples in interference pattern. The linear asymmetric ripples are gradational with sinuous ripples in the middleground and the sinuous ripples are gradational with cuspate ripples in the background. Length of debris in center is about 45 cm.

Fig.59. Complex interference ripple marks, Cliff Creek. Cuspate ripples are extensively modified by secondary ripples. Main channel is at the bottom, bank is to the top. Currents flowing from left to right produced the cuspate ripples. These were later modified by waves lapping up on the bank. Length of scale is 15 cm.

## COMPLEX INTERFERENCE RIPPLE MARK: LINEAR ASYMMETRIC AND SECONDARY, CUSPATE AND SECONDARY

(compound ripple mark, cross-ripple, dimpled current mark, tadpole nest)

### Descriptions

Complex interference ripple marks are formed from sets of more than one type of ripple. Along ephemeral streams in the Uinta Basin secondary ripples combine

Fig.60. Complex interference ripple marks formed by one set of linear asymmetric ripples and multiple sets of secondary ripples in channel margin, Dripping Rock Creek. Note that the relief of the linear ripples is partially controlled by the secondary ripples, indicating sequential development. Main flow in channel is from right to left. The linear ripples are asymmetric upstream, indicating a local backwater effect. Scale is 15 cm.

with linear asymmetric and cuspate ripples to form complex interference patterns (Fig.58–60). Within each interference pattern one set of ripples is dominant. Generally the cuspate or linear ripples are strongly developed and only partially modified by secondary ripples. Rarely, modification by secondary ripples is so complete as to almost obscure the large ripples. The angle between the two ripple sets is variable, but usually is large (Fig.61–63). The overall pattern is rectangular. Because the larger ripples usually dominate, the interference pattern could be described as ladderback.

## Occurrence

Complex interference ripple marks are minor structures along ephemeral streams. They occur along the margins of channels and on bars. Because so few examples were found, we cannot generalize their mode of occurrence.

Cuspate and secondary ripple marks

| | | | | |
|---|---|---|---|---|
| – t – t – t t – – – – – | – r – – | – – – – | – – – – – – | – – – – |

Linear and secondary ripple marks

| | | | | |
|---|---|---|---|---|
| – – – t – t t – – – – – | t t – – | – – – – | – – – – – – | – – – – |

## Origin

In contrast to simple interference ripple marks, complex patterns form sequentially. First the linear or cuspate ripples are built, then they are modified by secondary

Fig.61. Cuspate ripple marks in main channel of Dripping Rock Creek. Dominant current is from left to right. The crests of the secondary ripple marks in the lower half of the picture are oriented nearly at right angles to the cuspate ripples.

Fig.62. Interference ripple marks formed by cuspate and secondary ripple marks. Secondary ripples are largely confined to troughs of cuspate ripples. Dominant downstream current from top to bottom of picture. Refracted current from left to right. Quarter for scale.

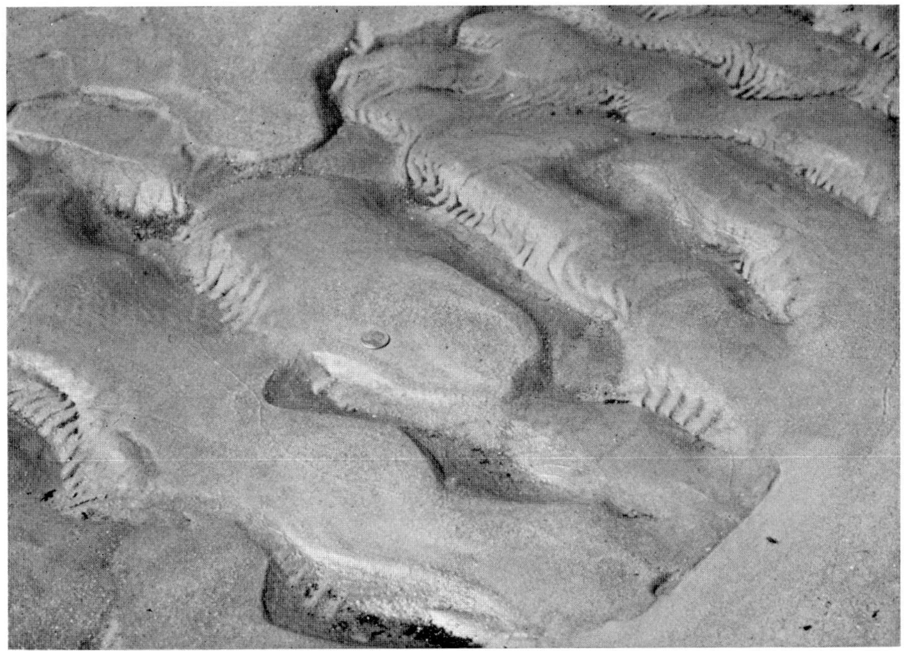

Fig.63. Complex interference ripples, Ashley Creek. Upstream from Fig.62. Note flatness of crests of cuspate ripples. Downstream current from top to bottom; secondary refracted waves from left to right. Quarter for scale.

ripples that form above the earlier ripples. A two-stage origin is shown by the continuity of secondary ripples across the crests of cuspate or linear ripples, the superposition of the two sets, and direct observation of their formation (Picard and High, 1970a). As a flood wanes, cuspate ripples, formed by low-velocity currents, or linear ripples, formed by refracted current-generated waves, may be stranded in quiet water behind or downstream of still-flooded bars. Secondary ripples may form to modify the pre-existing rippled bottom. Thus, complex interference ripples record the changing hydrodynamic conditions at a site during different stages as a flood recedes.

## Preservation and occurrence in sedimentary rocks

Interference ripple marks are common in the fossil record. Although the majority of them are simple interference patterns (two sets of overlapping wave ripples), complex sets similar to those described here are known. However, it is doubtful if complex interference ripple sets are often preserved in fluvial, or at least ephemeral fluvial, environments because the structures apparently are formed so rarely.

## Other occurrences

Complex interference ripple sets are common in the upper Berea Sandstone (Mississippian), a transgressive, shallow-marine deposit in Ohio. Similar features have also been reported from recent and ancient tidal flats (Van Straaten, 1954; Trefethen and Dow, 1960; Reineck, 1963; Klein, 1970; Wunderlich, 1970).

## Significance

Complex interference ripple marks are not diagnostic of fluvial environments. If found in fluvial rocks they are suggestive of bars or channel margins and are indicative of fluctuating hydrodynamic conditions in either tidal or ephemeral stream settings. They are a doubtful paleocurrent indicator.

## Selected references

Bucher, W. H., 1919. On ripples and related sedimentary surface forms and their paleogeographic interpretation. *Am. J. Sci.*, 4(47): 149–210; 241–269.

Klein, G. de V., 1970. Depositional and dispersal dynamics of intertidal sand bars. *J. Sediment. Petrol.*, 40: 1095–1127.

Picard, M. D. and High, L. R., Jr., 1970a. Interference ripple marks formed by ephemeral streams. *J. Sediment. Petrol.*, 40: 708–711.

Tanner, W. F., 1960. Shallow-water ripple mark varieties. *J. Sediment. Petrol.*, 30: 481–485.

Reineck, H.-E., 1963. Sedimentgefüge im Bereich der südlichen Nordsee. *Abhl. Senckenberg. Naturforsch. Ges.*, 505: 1–136.

Van Straaten, L. M. J. U., 1954. Sedimentology of recent tidal-flat deposits and the Psammites du Condroz (Devonian). *Geol. Mijnbouw*, 16: 25–47.

## STREAMING LINEATION

(sand streak, parting lineation, primary current lineation, parting plane lineation, harrow mark, graination)

### Description

Streaming lineation was first described by Sorby (1908, p.180). The structure has since been confused with parting lineation (Allen, 1964). McBride and Yeakel (1963) differentiated the two structures as parting plane lineation (streaming lineation) and parting step lineation (parting lineation). The name streaming lineation was given to ancient and modern examples by Conybeare and Crook (1968, p.25). We prefer the term streaming lineation, rather than parting plane lineation, to clearly distinguish this structure from parting lineation.

Streaming lineation is a series of low relief, sub-parallel ridges and grooves that cover bedding surfaces. Individual ridges and grooves are generally less than 3 cm in width, less than 0.5 cm in height, and up to about 1 m in length. The marked surface may be only several square meters in area or it may cover hundreds of square meters. Streaming lineation is oriented parallel with the channel (Fig.64, 65).

### Occurrence

Streaming lineation is one of the major sedimentary structures found along ephe-

Streaming lineation

| r r r c r c c c r c a r | t c a a | a a a a | – – a a – | – r – – |
|---|---|---|---|---|

meral streams. The structure is widespread on channel floors and on the tops of transverse and longitudinal bars. Locally, streaming lineation is developed almost to the exclusion of other structures. Associated structures are lag ripples, sinuous ripples and crescent scours. Streaming lineation is best developed in coarse gravelly sand and is most abundant along the upper reaches of streams.

### Origin

Streaming lineation is formed by relatively high-velocity currents. Upper-flow regime conditions are indicated by planar bedding and washed-out ripples. For

Fig.64. Streaming lineation on channel floor, Little Mountain. Flow is from left to right over toe of longitudinal bar. Note development of lag ripples to right in foreground. Length of scale is 15 cm.

Fig.65. Streaming lineation and crescent scours, Dripping Rock Creek. Flow is toward viewer. Hammer for scale.

coarse sand, current velocities in excess of 60 cm/sec probably are required. Streaming lineation represents the highest velocity bedform we have found in large abundance along ephemeral streams. Small amounts of antidune bedding were observed.

The association of streaming lineation with cuspate and secondary ripples demonstrates the relative current velocities in effect during formation of each structure. Just as the crests of cuspate ripples are drawn downstream to form sinuous ripples, so are the crests of sinuous ripples drawn downstream to form streaming lineation (Fig.66, 67). Thus, the sequence from cuspate ripples to sinuous ripples to streaming lineation represents a progressive increase in current velocity.

**Preservation and occurrence in sedimentary rocks**

Streaming lineation has been recognized in only a few rocks (Conybeare and Crook, 1968, p.25). However, many reported occurrences of parting lineation may in fact be streaming lineation. In ephemeral streams, streaming lineation is one of the few structures that is likely to be commonly preserved. Trenches cut into

Fig.66. Streaming lineation, sinuous ripples and cuspate ripples, Twelvemile Wash. Banks of channel in upper right and left center. Streaming lineation in thalweg. Cuspate ripples on longitudinal bar. Note forward deflection of ripple crests in transition from one structure to next. Flow is toward viewer. Hammer for scale.

Fig.67. Streaming lineation (top), sinuous ripples (center) and cuspate ripples (bottom), Twelve-mile Wash. Flow is from left to right. The relative current velocities are indicated by the orientation of the ripple crests. Hammer for scale.

channel deposits show that planar bedding is the most abundant type. The relative abundance of planar bedding arises from the nature of deposition in ephemeral streams. During the waxing stage of floods, erosion is dominant and a relatively deep channel is cut. As the flood wanes, the channel is filled with sediment. Initially, upper-flow regime conditions persist and the bulk of the channel fill is deposited as plane beds, probably marked by streaming lineation. Only during the late stages of flooding do lower-flow regime conditions exist, producing extensive rippled surfaces in the upper few centimeters of channel fills. Thus, the mode of formation of streaming lineation insures the preservation of the structure. In contrast, ripple marks and other structures that are formed only on exposed channel floors and not throughout channel fill sequences are less likely to be buried and preserved.

**Other occurrences**

Streaming lineation is also formed on beaches (Conybeare and Crook, 1968, p.25). In view of the high current velocities required for its formation, streaming lineation is probably restricted to ephemeral streams, the swash zone of beaches, and some tidal settings (M. O. Hayes, personal communication, 1972).

### Significance

Streaming lineation is not diagnostic of fluvial environments. If present in fluvial rocks, it is probably indicative of ephemeral flow, and is restricted to channel subenvironments. Transitional or upper-flow regime conditions are necessary for the formation of streaming lineation. The structure is formed parallel with the direction of current flow.

### Selected references

Allen, J. R. L., 1964. Primary current lineation in the Lower Old Red Sandstone (Devonian), Anglo-Welsh Basin. *Sedimentology*, 3: 89–108.
Conybeare, C. E. B. and Crook, K. A. W., 1968. Manual of sedimentary structures. *Bull. Bur. Min. Resources*, 102: 327 pp.
Karcz, I., 1966. Secondary currents and the configuration of a natural stream bed. *J. Geophys. Res.*, 81: 3109–3112.
Karcz, I., 1967. Harrow marks, current-aligned sedimentary structures. *J. Geol.*, 75: 113–121.
McBride, E. F. and Yeakel, L. S., 1963. Relationship between parting lineation and rock fabric. *J. Sediment. Petrol.*, 33: 779–782.
Sorby, H. C., 1908. On the application of quantitative methods to the study of the structure and history of rocks. *Q. J. Geol. Soc. Lond.*, 64: 171–232.

## PARTING LINEATION

(primary current lineation, current lineation, current parting, parting step lineation)

### Description

Parting lineation is a series of small, sub-parallel, shallow ridges and grooves developed on parting planes in thinly laminated sediment. Relief between ridges and grooves is slight, generally only a few hundredths of a centimeter. The widths of grooves and ridges is very much greater than the height and both are flat. The overall appearance is of low, parallel steps separated by relatively broad, flat areas (Fig.68, 69).

### Occurrence

Because parting lineation is only expressed on parting planes, and not on the surface, the actual abundance of the structure may be much greater than is indi-

Fig.68. Parting lineation on channel floor, Cliff Creek. Overlying mud cracked sediment has been removed. Flow is from right to left. Scale is 15 cm.

Fig.69. Parting lineation on channel floor, Cliff Creek. Scale is 15 cm.

Fig.70. Parting lineation on side of channel, Dripping Rock Creek. Flow in main channel (to right) is from top to bottom. Parting lineation is in mud plastered on channel bank where water drained into channel. Pen is 13.5 cm long.

Parting lineation

| c – t – t – – t t t – – | a – r r | – r r – | a a – r – | a – a a |
|---|---|---|---|---|

cated here. Parting lineation occurs on point bars and broad, open channels that are extensively mud cracked (Fig.70, 71). Where the surficial layer has peeled back, parting lineation commonly is visible on the underlying fine sand. The structure is most abundant in the channel along lower reaches of ephemeral streams; upstream it is only found on point bars. Parting lineation commonly occurs in fining-upward

Fig.71. Parting lineation on channel floor, near mouth of Twelvemile Wash. Flow is from left to right. Note the alignment of the mud curls, which reflects the anisotropic grain fabric. Scale (center) is 15 cm.

sequences above micro cross-lamination (cuspate ripples) and below mud cracks.

## Origin

Parting lineation is a secondary effect of subparallel grain fabric in laminated deposits. As the thin beds are separated, splitting breaks across several bedding planes. Because of the oriented internal fabric of each layer, these layer-to-layer breaks are easiest in one direction, parallel with the long axes of grains. With random grain orientation, splitting produces an irregular pattern rather than the subdued, oriented steps of parting lineation. Thus, the steps of parting lineation are parallel with the long axes of oriented individual sand grains. This relationship between structure and grain fabric has been demonstrated by McBride and Yeakel (1963, pp.779–782) and Picard and Hulen (1969, pp.2631–2636) and the origin of parting lineation is to be found in the origin of subparallel grain fabric.

Individual grains are oriented by current action. If the grains are markedly non-equidimensional the long axes are rotated so that they are parallel with current flow. This streamlined position is the most stable configuration and given sufficient current action, it will be assumed by most grains. However, the intensity of the current is a disputed subject. Parting lineation has been considered to form under conditions of low flow velocity (Sorby, 1908, p.180), moderate velocity (Stokes, 1947, p.54), and high velocity (Sorby, 1908, p.180; Allen, 1964, pp.101–106; Picard and Hulen, 1969, p.2635). In part, this confusion has resulted from the misidentification of streaming lineation, an upper-flow regime structure, with parting lineation. Our information indicates that parting lineation is formed by low velocity currents. Observations in support of this interpretation are: (*1*) relatively fine grain size compared with that in adjacent ripples; (*2*) association with mud cracks and suspension deposits; (*3*) position in fining upward sequences (waning flood) above cuspate ripples; (*4*) distal position along ephemeral stream channels; and (*5*) distinction from streaming lineation. Upper-flow regime conditions are probable but not certain. Despite the decreased flow velocity relative to ripple marks, the fine grain-size and shallow-water over bars may still result in upper-flow regime conditions. All modern flume experiments indicate that plane beds are upper-flow regime structures and no transportation occurs below the ripple field. Nevertheless, plane beds with surface graining were produced by Sorby (1908, p.180) by currents insufficient to cause ripples and the effects of low-velocity currents on fine and medium sand deserve additional study. The common sequence of ripple stratification, plane beds with parting lineation, and suspension deposits records waning floods and low-flow velocities. If parting lineation is an upper-flow regime structure, then conditions went immediately from upper-flow regime to quiet enough to allow suspended grains to settle; the absolute lack of intervening ripples indicates that the lower-flow regime was somehow bypassed. This dilemma cannot be resolved without further study of the structure.

**Preservation and occurrence in sedimentary rocks**

In contrast to most sedimentary structures, parting lineation is common in fluvial sandstone. In addition, the structure is widely distributed in thinly bedded sandstone of other environments. The reason for the relative abundance of parting lineation is that it is the product of the internal fabric, rather than just a surface marking. Thus, the structure, once produced, is buried and protected. Furthermore, it is volumetrically abundant, being distributed throughout a deposit rather than just here and there on the surface.

**Other occurrences**

The structure occurs widely in rocks of fluvial, beach, shallow marine and turbidite environments.

### Significance

Parting lineation is not diagnostic of fluvial environments or subenvironments. It is formed parallel with the direction of current flow under lower-flow regime conditions. Characteristically, parting lineation is formed at the distal ends of ephemeral streams.

### Selected references

Allen, J. R. L., 1964. Primary current lineation in the Lower Old Red Sandstone (Devonian), Anglo-Welsh Basin. *Sedimentology*, 3: 89–108.
McBride, E. F. and Yeakel, L. S., 1963. Relationship between parting lineation and rock fabric. *J. Sediment. Petrol.*, 33: 779–782.
Picard, M. D. and Hulen, J. B., 1969. Parting lineation in siltstone. *Bull. Geol. Soc. Am.*, 80: 2631–2636.
Sorby, H. C., 1908. On the application of quantitative methods to the study of the structure and history of rocks: *Q. J. Geol. Soc. Lond.*, 64: 171–232.

## IMBRICATION
(edgewise structure, shingle structure)

### Description

Imbrication is the preferred orientation of flat pebbles that are deposited slightly overlapping one another and dipping upstream. The appearance is that of shingles. Although the structure is best developed in platy gravel, imbrication also occurs in ellipsoidal pebbles (Fig.72, 73).

### Occurrence

Imbrication is a relatively minor structure along ephemeral streams in the Uinta

Imbrication

| | | | | |
|---|---|---|---|---|
| t r r c t r – – r – – – | – – r c | a r – – | – – – – – | c c t – |

Basin. It is most abundant along the upper reaches of streams where the gradient is high and the load composed of gravel. To be extensively developed, the structure requires large amounts of gravel; individual pebbles must overlap. If lesser amounts of gravel are present, imbrication only occurs in scattered patches where the coarse sediment has been concentrated. If size sorting has not concentrated the gravel, crescent scours develop in place of imbrication.

Fig.72. Imbrication of flat pebbles on stream bed, Dripping Rock Creek. View looking upstream. When seen from this position, only the edges of flat pebbles are visible; when viewed downstream, the flat tops are apparent. Dimensions of the notebook are 12.5 by 20 cm.

## Origin

Imbrication is formed by current action. The upstream-dipping shingle orientation of flat pebbles is the most stable configuration. In this position currents are deflected off the flat top of the pebble and are unable to get underneath and flip the pebble over. No estimate of current velocity has been made.

Fig.73. Imbrication of ellipsoidal pebbles exposed in stream bank, Halfway Hollow. Flow was from left to right. Hammer for scale.

**Preservation and occurrence in sedimentary rocks**

In common with streaming lineation and parting lineation, imbrication is an internal structure and not a surface marking. Consequently, preservation of the structure is greatly enhanced. Imbrication is a fairly common feature of conglomerates.

**Other occurrences**

Imbrication occurs wherever gravel is concentrated and is subject to current action. In addition to all streams, imbrication occurs along beaches and tidal flats.

**Significance**

Imbrication is not diagnostic of fluvial environments. In fluvial rocks it is indicative of channel conditions and pebbles dip upstream.

## Selected references

Cailleux, A., 1945. Distinction des galets marins et fluviatiles: *Bull. Soc. Géol. France*, 15: 375–404.
Fraser, H. J., 1935. Experimental study of the porosity and permeability of clastic sediments. *J. Geol.*, 43: 910–1010.
Krumbein, W. C., 1939. Preferred orientation of pebbles in sedimentary deposits. *J. Geol.*, 47: 673–706.
Krumbein, W. C., 1940. Flood gravel of San Gabriel canyon. *Bull. Geol. Soc. Am.*, 51: 636–676.
Krumbein, W. C., 1942. Flood deposits of Arroyo Seco, Los Angeles County, California. *Bull. Geol. Soc. Am.*, 53: 1355–1402.
Lane, E. W. and Carlson, E. J., 1954. Some observations of the effect of particle shape on movement of coarse sediments. *Trans. Am. Geophys. Union*, 35: 453–462.
Potter, P. E. and Pettijohn, F. J., 1963. *Paleocurrents and Basin Analysis*. Academic Press, New York, N.Y., 296 pp.
Schlee, J., 1957. Fluvial gravel fabric. *J. Sediment. Petrol.*, 27: 162–176.
White, W. S., 1952. Imbrication and initial dip in a Keweenawan conglomerate bed. *J. Sediment. Petrol.*, 22: 189–199.

*Chapter 4*

# POST-DEPOSITIONAL STRUCTURES

The literature on post-depositional structures is variable in terms of studies of individual structures. Although a few structures, such as slump features, have been extensively studied, many questions even remain concerning the origin and occurrence of slump features. For other, less well known structures, the problems are more severe. A notable example is shrinkage cracks. Although this structure is

TABLE XI

Classification of post-depositional structures

| *Time of formation* | *Origin* | *Feature* |
|---|---|---|
| Pre-burial | desiccation | polygonal shrinkage crack |
| | | linear-shrinkage crack |
| | | mud curl |
| | | mud pebble |
| | | reverse mud curl |
| | | earth crack |
| | saline crystallization | salt crust |
| | | salt ridge |
| | | wrinkled surface |
| | | crystal mold |
| | | (reverse mud curl) |
| | external activity | track, trail, burrow |
| | | raindrop impression |
| | | textured surface |
| | biologic activity | algal mat |
| | | algal mound |
| Post-burial | compaction | gas bubble |
| | | sand volcano |
| | | convolute bedding* |
| | | flame structure* |
| | | bioturbation* |

* See Chapter 5.

widespread in both modern and ancient deposits, various morphologic varieties have rarely been differentiated.

Post-depositional structures result from events that take place after sediment is deposited. These structures include features that are intrinsic to the sediment (desiccation, compaction, salt crust structures) and those that are formed accidentally (tracks, trails, various surface markings). Algal-formed features associated with ephemeral streams are also included in this chapter.

Except for compactional features (gas bubbles, sand volcanoes), the post-depositional structures described here are those that form prior to burial. The abundance of post-burial structures that are not expressed at the surface (such as convolute bedding) could not be estimated. Accordingly, these structures (disturbed bedding) are considered in the next chapter with bedding types.

The classification of post-depositional structures shown in Table XI is not intended as a general classification of these structures. Rather, we have grouped related features that tend to occur together along ephemeral streams.

The abundance of post-depositional structures defines two general populations: those whose average abundance is rare to common (polygonal shrinkage crack, mud curl, earth crack, tracks, raindrop impression, and algal mound) and those that occur in trace amounts (linear-shrinkage crack, mud pebble, reverse mud curl, salt crust, salt ridge, wrinkled surface, salt crystal, gas bubble, sand volcano, and algal mat). Only textured surfaces are intermediate between these two groups (trace to rare in abundance).

## POLYGONAL SHRINKAGE CRACK
(mud crack, desiccation crack, sun crack)

### Description

Polygonal shrinkage cracks are the familiar variety of mud cracks. Four-sided patterns are the most common but three- to six-sided forms occur. At several localities along Cliff Creek about three-fourths of the polygons were four-sided; of the remaining polygons, three-sided polygons were twice as abundant as five-sided polygons. Six-sided forms apparently are rare and are the least abundant of the polygonal shrinkage cracks. Size ranges from less than 3 cm along a side to more than 30 cm. Depth of cracking is roughly related to the length of cracks and reaches a maximum of 5 cm.

Several varieties of polygonal cracks are formed along ephemeral streams. The most common type is characterized by random cracks. No single direction is preferred over other possible directions (Fig.74,75). A common modification exhibits cracks in one direction (generally parallel with the channel) that are better developed than associated cross-cracks (Fig.76). A third variety is characterized by major cracks that are oriented into arcs or swirls (Fig.77).

## Occurrence

Polygonal shrinkage cracks are distributed widely along most ephemeral streams.

Polygonal shrinkage crack

| | | | | |
|---|---|---|---|---|
| a c r c a c c r c r – t | a c r r | c r c t | a a c a r | a c a a |

Cracks occur with equal abundance within channels, on bars and on floodplains. Cracks in the channel are most abundant downstream where they cover extensive surfaces. Upstream, cracks are restricted to small hollows and scour holes. Cracks develop only in fine-grained sediment.

Polygonal cracks whose longer sides develop parallel with the channel are formed only within the channel. Arcuate crack patterns are developed on floodplains or downstream of bars.

Fig.74. Polygonal shrinkage cracks in scour hole, Cliff Creek. Four- and five-sided polygons are dominant. Large size of cracks results from thick mud layer. Trowel for scale (right center).

Fig.75. Polygonal shrinkage cracks on floodplain, Dripping Rock Creek. Three- and four-sided forms are dominant.

Fig.76. Oriented polygonal shrinkage cracks in channel, Cliff Creek. Flow is toward viewer. Cracks are the result of subparallel grain fabric. Shovel for scale (right center).

Fig.77. Arcuate shrinkage crack pattern on floodplain, Cliff Creek. Photograph is from an area adjacent to Fig.76. Hammer for scale.

## Origin

Shrinkage cracks result from a reduction in volume as fine-grained sediment dries if exposed to the air, or compacts, if under standing water. Ideally, it has been suggested that in homogeneous material cracks radiate from evenly spaced points at 60°-angles and the intersections of the cracks form six-sided polygons (internal angles equal 60°). However, in drying mud layers it is our experience that this "ideal" is rarely achieved and most angles approximate 90°, resulting in four-sided forms. There are some experimental results that support our field experience. Anderson and Everett (1964, p.5), using varied sand and silt-fractions spread on wet sand, found that, ". . . primary cracks generally originate as single linear features and propagate along straight or slightly curved lines until they encounter an inhomogeneity in the texture or stress field of the mud flat. At such inhomogeneities the cracks commonly bifurcate. Secondary cracks branch off from primary cracks, generally at right angles, and then propagate like primary cracks, or else originate as simple linear features between primary cracks. Regardless of their original direction of propagation, secondary cracks curve to join previously formed cracks at nearly right angles."

Fig.78. Close-up of arcuate cracks in Fig.77 showing control that is exerted by the orientation of plant remains. Juniper needles and twigs were incorporated in a froth that floated on top of the flood waters. Swirling patterns in the froth were produced by backwater eddies. After flood recession the froth was stranded on the floodplain.

Cracks that show preferred orientation reflect anisotropic grain fabric within the drying layer of sediment. For cracks oriented parallel with the channel, the grain orientation is produced by low-velocity currents during final flood stages as indicated by the associated parting lineation. For arcuate and swirling patterns, the cracks parallel the fabric of organic matter (juniper needles, twigs, and so forth) in the mud (Fig.78). Plant remains are commonly incorporated in a froth that floats on top of turbid flood waters. Rafting of the froth onto bars and floodplains, together with eddy motions, produces the distinctive organic mud layer and resulting crack pattern.

## Preservation and occurrence in sedimentary rocks

Shrinkage cracks are common structures in sedimentary rocks. Along ephemeral streams the likelihood of preservation is good, reflecting the abundance of the structure and their penetration into the sediment.

### Other occurrences

Shrinkage cracks are distributed widely in all fluvial settings. In addition, the structure is common in lacustrine, tidal and shoreline environments.

### Significance

Polygonal shrinkage cracks are not diagnostic of fluvial environments or subenvironments. If found in channel beds, they are indicative of ephemeral flow conditions. In some occurrences they can be used to determine the current direction. They are generally indicative of exposure to the atmosphere, but can originate in very shallow water settings.

### Selected references

Bradley, W. H., 1933. Factors that determine the curvature of mud-cracked layers. *Am. J. Sci.*, 26: 55–71.

Karcz, I. and Goldberg, M., 1967. Ripple controlled desiccation patterns from Wadi Shiqma, southern Israel. *J. Sediment. Petrol.*, 37: 1244–1245.

Lachenbruch, A. H., 1961. Depth and spacing of tension cracks. *Geophys. Res.*, 66: 4273–4292.

Longwell, C. R., 1928. Three common types of desert mud cracks. *Am. J. Sci.*, 15: 136–145.

## LINEAR-SHRINKAGE CRACK

(incomplete shrinkage cracks, pseudo mud crack, radiate mud crack, subaqueous shrinkage crack)

### Description

Linear-shrinkage cracks are open, straight to curved cracks that occur singly or in sets. Individual cracks are 2–8 cm in length; bifurcations are rare. Many linear cracks show preferred orientation, either parallel with the sides of depressions or parallel with the crests of ripple marks (Fig.79–82).

### Occurrence

Linear-shrinkage cracks are largely restricted to channels.

Linear cracks

| r t – – – – t – – – – – | r – – – | – t – – | r r r – – | t c – t |
|---|---|---|---|---|

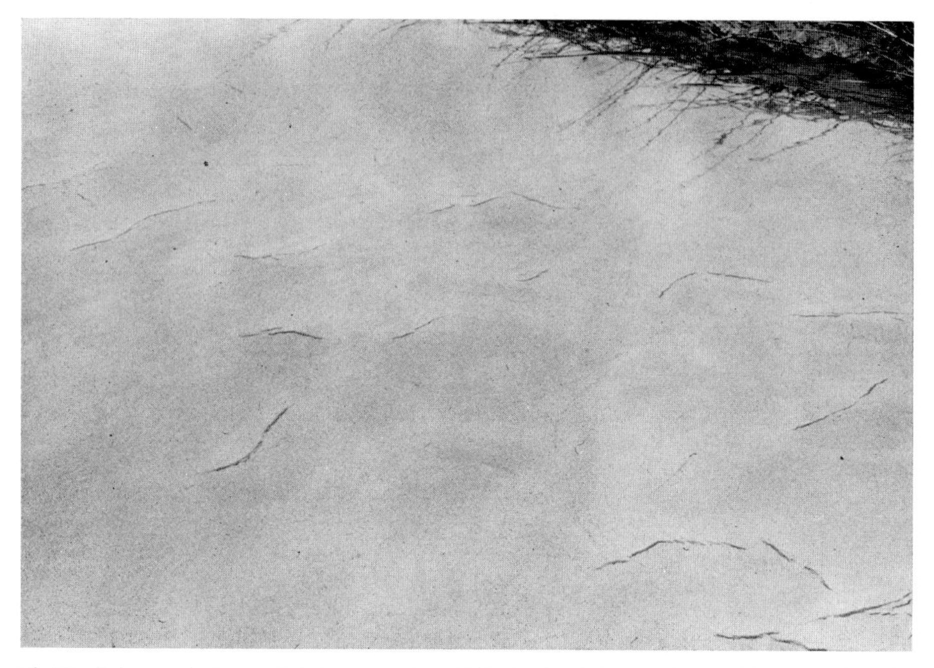

Fig.79. Submerged, linear-shrinkage cracks, Cliff Creek. Subparallel trend is approximately perpendicular to stream bank. Maximum length is about 30 cm. Cracks in mud film over sand.

Their distribution is spotty; overall they are a minor structure but locally they are abundant. There is no apparent pattern to the distribution of this structure along the lengths of streams. Structures commonly associated with linear cracks include other post-depositional surface markings and cuspate ripple marks.

## Origin

Linear-shrinkage cracks have been ascribed to a variety of origins, both inorganic and organic. Markedly curved features have been considered to be the tracks of animals (Faul, 1950, p.102; Frarey and McLaren, 1963; Frarey et al., 1963). However, the organic origin of these features has been received with skepticism (Häntzschel, 1949; Wheeler and Quinlan, 1951, p.141; Barnes and Smith, 1964; Cloud, 1968, pp.29–32). Laporte (1968, p.15) suggested that linear-shrinkage cracks are indicative of formation under standing water.

Analysis of ancient and modern linear cracks has led us to conclude that the dominant control is topography (Picard, 1966,1969; Picard and High, 1969). Linear cracks develop when relatively thick, water-saturated thixotropic muds dehydrate, usually under standing water. If the surface is exposed, evaporation is more rapid and smaller polygonal cracks tend to develop in surficial layers.

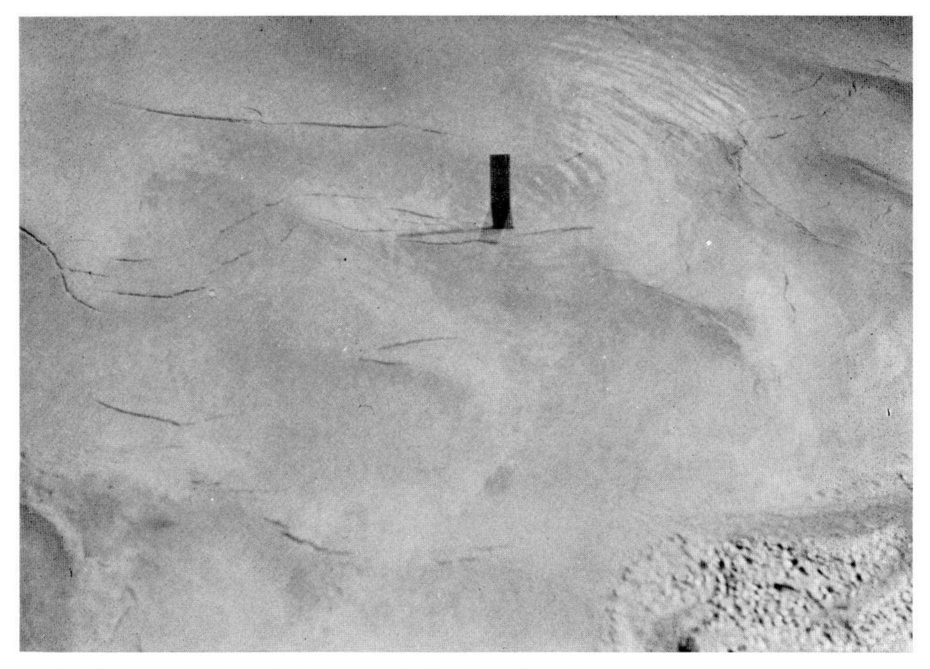

Fig.80. Submerged, linear-shrinkage cracks that have formed near crests of cuspate ripple marks, Cliff Creek. Exposed length of scale is 12.5 cm.

However, if the surface is submerged, shrinkage is slow and deep linear cracks form. Where the bottom is irregular, slight downslope gravity movement imparts an orientation to the linear cracks. Around the margins of depressions, cracks are circumferential. In mud draped over ripple marks, the cracks tend to parallel ripple mark crests.

**Preservation and occurrence in sedimentary rocks**

Linear-shrinkage cracks are the most abundant bedding plane structure in lacustrine near-shore siltstone, sandstone and carbonate of the Parachute Creek Member of the Green River Formation (Eocene). The structure has also been found in several marine or marginal marine formations. Preservation in ephemeral streams is possible but unlikely. The structure is developed on channel floors, where it is subject to weathering while exposed or to erosion by subsequent floods.

**Other occurrences**

In rocks, linear-shrinkage cracks have been reported from shallow lacustrine, marine and marginal marine beds.

Fig.81. Linear-shrinkage cracks in mud film over cuspate ripples in channel near bank of Cliff Creek. Current is from left to right. Rain imprints are visible. Maximum length is about 30 cm (for long dimension of crack in left bottom).

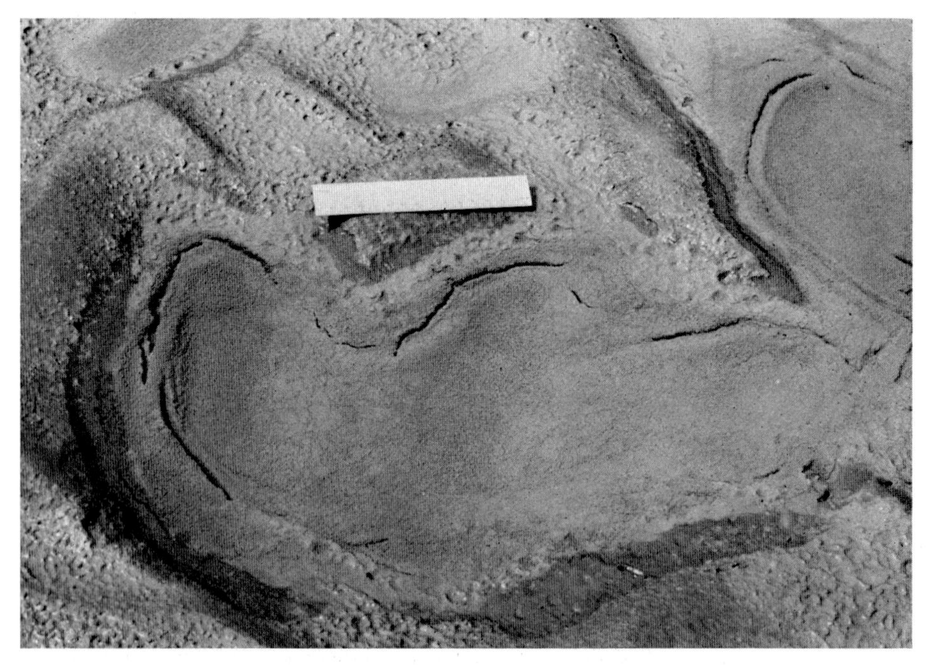

Fig.82. Exposed, curved and linear-shrinkage cracks bounding smaller, polygonal shrinkage cracks (center of photograph) at Cliff Creek. Length of scale is 15 cm.

## Significance

Linear-shrinkage cracks are not indicative of fluvial environments. If present in fluvial beds, they can be used to determine the paleocurrent direction, being generally parallel to stream flow. However, linear cracks can also form over local relief features, such as ripples or depressions, and be unrelated to stream direction.

## Selected references

Burst, J. F., 1965. Subaqueously formed shrinkage cracks in clay. *J. Sediment. Petrol.*, 35: 348–353.
Jüngst, H., 1934. Geological significance of synaeresis. *Geol. Rundsch.*, 25: 321–325.
Picard, M. D., 1966. Oriented, linear-shrinkage cracks in Green River Formation (Eocene), Raven Ridge area, Uinta Basin, Utah. *J. Sediment. Petrol.*, 36: 1050–1057.
Picard, M. D., 1969. Oriented, linear-shrinkage cracks in Alcova Limestone Member (Triassic), southeastern Wyoming. *Contrib. Geol.*, 8: 1–7.
Picard, M. D. and High, L. R., Jr., 1969. Some sedimentary structures resulting from flash floods. *Bull. Utah Geol. Mineral. Surv.*, 82: 175–190.
Van Houten, F. B., 1964. Cyclic lacustrine sedimentation, Upper Triassic Lockatong Formation, central New Jersey and adjacent areas. *Kansas Geol. Surv. Bull.*, 169: 497–531.
White, W. A., 1961. Colloid phenomena in sedimentation of argillaceous rocks. *J. Sediment. Petrol.*, 31: 560–570.

# MUD CURL
(clay gall, shale pebble, mud crust)
MUD PEBBLE

## Description

Mud curls are thin layers of fine grained sediment that have curved upward during dewatering of shrinkage polygons. Commonly, curls are so strongly curved that they form cylindrical rolls (Fig.83). The thickness of the curled mud layer is about 0.2 cm and polygon widths range from about 5 to 15 cm.

## Occurrence

Mud curls are widely distributed along ephemeral streams and locally are the most

Mud curl

| r c r c r r c r r t – – | a c r t | r t c t | c a a c t | c c c a |
|---|---|---|---|---|

abundant structure. Mud curls are most often found in channels although they

Fig.83. Mud curls on point bar in Cliff Creek. Scale in middleground is 15 cm long. Flow from top to bottom.

occur on bars and on floodplains. The structure is most abundant along downstream reaches, in common with shrinkage cracks and parting lineation.

## Origin

Mud curls are dehydration shrinkage features that, in common with shrinkage polygons, result from the drying of surficial mud layers. The only difference in the two structures is the upward curvature of mud curls. Two factors probably explain the curling of mud polygons. As a thin mud layer (generally less than 2.5 cm) dries, surface evaporation is fast and dehydration at depth is progressively slower. Because shrinkage at the top is more rapid, the layer curls in that direction. Secondly, most mud layers fine upward. The finer grained parts of the layer contain more water than do the coarser basal parts. Consequently there is more shrinkage at the top than at the bottom and curling results.

## Preservation and occurrence in sedimentary rocks

Mud curls are delicate structures that are not commonly preserved. However,

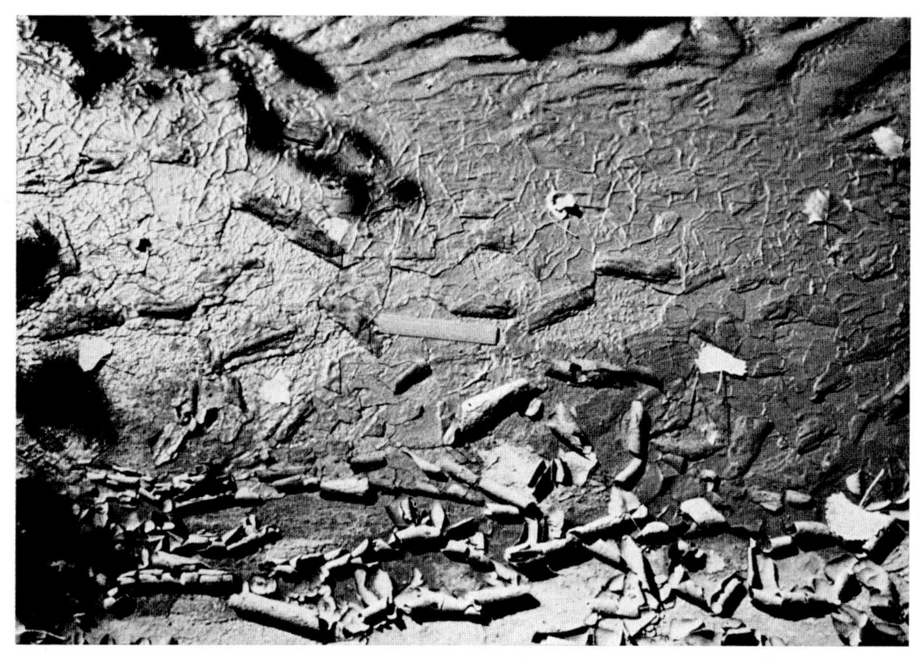

Fig.84. Mud pebbles on point bar, Twelvemile Wash. Note rewetted polygons and slight distance of transport. Flow is from left to right. Cuspate ripples in channel at top; mud curls and parting lineation on bar at bottom. Length of scale is 15 cm.

preservation of fragments is more likely. Subsequent floods break up the mud curls and transport the fragments. If transportation is not too turbulent or far, the fragments can survive to be redeposited (Fig.84). The relative abundance of mud pebble conglomerate in ancient beds indicates that preservation does occur. The relative scarcity of redeposited mud pebbles along ephemeral streams is probably

Mud pebble

| | | | | |
|---|---|---|---|---|
| – t – – – t – – – – – – | r – – – | – – – – | – r – t – | – – t – |

not indicative of their true abundance. Most mud pebbles would be buried within the sediment and would not appear on the surface.

**Other occurrences**

Mud curls are widespread in most terrestrial environments and on high tidal flats. Mud pebbles are common in fluvial, tidal and shallow marine rocks and in turbidites.

**Significance**

Mud curls are not diagnostic of fluvial environments. If present in fluvial rocks they are suggestive of floodplains or ephemeral channels. Mud pebbles indicate the presence of local currents and short distances of transport. Mud curls are indicative of subaerial exposure.

## REVERSE MUD CURL

**Description**

The distinctive feature of reverse mud curls is the downward curvature, the opposite of mud curls. Reverse curls also differ in size and shape, being smaller than other shrinkage polygons and more irregular in outline. In contrast to the regular three- and four-sided polygons common to most mud cracks, reverse curls cannot be characterized simply. Each side is uneven and the number of sides is variable.

**Occurrence**

Reverse curls are associated with either scour holes or salt crusts. The occurrence in Cliff Creek demonstrates the association with saline minerals (Fig.85,86).

Reverse mud curl

| | | | | |
|---|---|---|---|---|
| r – – r – – – – t – – – | – – – – | – – – – | – c – t – | – – t – |

Above Cow Wash, which is the source of saline sediment, there are no reverse curls. Below Cow Wash, reverse curls are common. In scour holes reverse curls are restricted to the bottom where the mud is thickest. On the sides of scour holes, reverse curls are replaced by normal cracks (Fig.87). Reverse curls are present within the channel and on low channel bars.

**Origin**

The direction of curvature of drying mud layers was debated by several workers early in this century. Shrock (1948, p.203) has reviewed the major papers. The initial view was that concave up polygons were deposited in fresh water (Barrell, 1913, p.459; Kindle,1917, pp.135–144) and concave down polygons in saline water (Kindle,1917, pp.135–144). This simple relationship between salinity and curvature has been shown to be incorrect by Ward (1923, pp.308, 309), Twenhofel (1932,

Fig.85. Reverse curls on low longitudinal bar, Cliff Creek opposite mouth of Cow Wash. The outlines of larger, pre-existing shrinkage polygons are visible filled with saline minerals. Length of scale is 15 cm.

pp.686–690), Bradley (1933) and Dow (1964). Shrock's photograph (1948, p.203) of mud cracks along Kanab Creek, Arizona, supports his conclusion that flat, concave and convex polygons can all be formed essentially simultaneously on the same surface in the same mud and water. Minter (1970, pp.755, 756) cites deposition on slight topographic highs where subsurface drainage below the mud film exceeded surface evaporation as the cause of downward curving mud polygons.

The reverse mud curls we observed do not support Minter's conclusions. Rather, multiple origins are required for this structure.

The observed gradation from mud curls to flat polygons to reverse curls to salt crust indicates a secondary origin with reverse curls forming from pre-existing shrinkage polygons and mud curls (Fig.85, 86). The first step in the formation of reverse curls is the deposition and dehydration of a mud layer, forming shrinkage polygons or mud curls. Continued evaporation at the surface draws water up from the substrate by capillary action. If this water is saline, a salt crust is formed. Evaporation is greatest on topographic highs and the salt crust is thickest on bars and thin to absent in the channel. Initial structures in the mud layer are destroyed as the salt crystals grow and the clastic grains are pushed apart. Reverse shrinkage

Fig.86. Reverse curls that are transitional between mud curls and salt crust, Cliff Creek. Mud curls at top are in the stream channel. Salt crust is on a lateral bar. Flat polygons and reverse curls are on the bar flank. Pen is 13.5 cm long.

cracks are intermediate, partially destroyed polygons. The polygons are first flattened by being broken into several pieces. Each fragment then forms domes or reverse curls because salt crystallization and expansion are more rapid at the surface. Continued salt growth disrupts the initial fabric and the mud layer crumbles, to be replaced by a salt crust. While original shrinkage polygons may remain unaltered within the channel, on the tops of bars a thick salt crust has replaced the mud layer. In between, the polygons are altered to reverse curls (Fig.86).

The formation of reverse curls in scour holes probably is related to thickness and character of the mud. Shrinkage polygons form around the margins of the

Fig.87. Reverse mud curls in scour hole, Kennedy Wash. Downward curvature results from draping of surficial layer over deeply cracked blocks in thick mud below. Normal mud curls on flank of scour at top. Shovel handle for scale.

scour hole where the mud layer is relatively thin. These cracks penetrate through the mud and the edges of the polygons turn up. However, in the center of the scour continued drying of the thick mud below the surficial layer causes surficial polygons to be draped over underlying blocks, resulting in a downward curvature. The failure of the thick mud in the bottom of the depression to form large, deeply penetrating cracks (Fig.74) may result from relatively rapid exposure and lack of standing water. Excessive shrinkage of the subsurface mud relative to the surficial layer may reflect its high organic content or its fine grain size.

### Preservation and occurrence in sedimentary rocks

The preservation of the rare reverse curls observed along Cliff Creek and Stinking Water Creek is unlikely. The sediment is soft and crumbly. The originally tough mud film has been disrupted by the growth of saline minerals and could not survive transportation. Preservation is possible only if the surface is buried quickly without scour. Reverse curls in scours are more likely to be preserved.

### Other occurrences

The reverse curls reported by Minter (1970, pp.755–756) and Shrock (1948, p.202) are not associated with saline deposits and must have a different origin, even though both are present along streams. Downward curving mud cracks have also been reported from marine settings.

### Significance

Reverse mud curls are not diagnostic of fluvial environments.

### Selected references

Shrock, R. R., 1948. *Sequence in Layered Rocks.* McGraw-Hill, New York, N.Y., 507 pp.

Minter, W. E. L., 1970. Origin of mud polygons that are concave downward. *J. Sediment. Petrol.*, 40: 755–756.

## EARTH CRACK
(subaerial sun-crack)

### Description

The final type of desiccation crack is characterized by irregular polygons formed in dry soil. The polygonal pattern is pronounced and four- and five-sided forms are the most abundant (Fig.88). In contrast to shrinkage polygons, the sides of earth cracks are irregular and not straight to curved. The sizes of individual polygons are small, generally measuring 5–15 cm, but the depth of cracking is relatively great, often exceeding 3 cm.

### Occurrence

Earth cracks are widespread on floodplains and on areas surrounding the stream. The structure was not observed within stream channels or on bars. Earth cracks

Fig.88. Earth cracks on terrace above Dripping Rock Creek. Note the irregularity of each side of the polygons. Small algal mounds are also present. Pen is 13.5 cm long.

| Earth crack | | | | |
|---|---|---|---|---|
| r a a a a c c a a c c a | r – c c | c c c c | c c c c c | a c c c |

develop in clayey soil that is thoroughly dry. The structure does not form in loose sand or in alluvium.

## Origin

Earth cracks are desiccation features of soils. The cracks open after the soil has been repeatedly wetted, by dew and showers, and then dried. The cracks are deep because the drying layer is thick, being the entire soil interval rather than a mud film.

## Preservation and occurrence in sedimentary rocks

Earth cracks are more resistent structures than are other types of desiccation cracks. In contrast to drying mud layers, earth cracks are at equilibrium with subaerial conditions and will remain on the surface for long periods. Thus, rapid burial is not essential. The greater depth of cracking also makes earth cracks more likely to be preserved. Preservation as vertical crack fillings or disturbed bedding should be more common than bedding surfaces.

### Other occurrences

Earth cracks are common on most land surface environments in subarid and arid climates.

### Significance

Earth cracks are indicative of fluvial regimes and are restricted to floodplains and uplands.

### Selected reference

Swartz, J. H., 1927. Subaerial sun-cracks. *Am. J. Sci.*, 14: 69–70.

## SALINE FEATURES: SALT CRUST, SALT RIDGE, WRINKLED SURFACE, CRYSTAL MOLD[1]

(creep wrinkles, crinkle marks, pseudo-ripples, "namak sefid")

### Description

Streams along which saline minerals are being formed show a variety of distinctive sedimentary features. The most typical feature is a salt crust (Fig.89). The salt crust has an irregular, knobby surface and reaches a maximum of about 2.5 cm in thickness. The crust is developed on a substrate of normal alluvium and is laterally extensive. Low-relief domes 4–8 cm in diameter are widespread in the salt crust. Similarly, low-relief ridges are common (Fig.90). The ridges form an irregular polygonal pattern and the mutual edges between adjacent polygons turn upward abruptly. The heights of ridges are 2–5 cm and most of them are broken along the crest.

Another structure associated with saline minerals is a finely wrinkled surface. A thin surficial mud film, less than 0.2 cm thick, is covered with small, low-relief wrinkles forming four- and five-sided polygons. The appearance is similar to the wrinkles that form on the film in scalded milk (Fig.91).

[1] Also see "Reverse mud curl" (p. 116.)

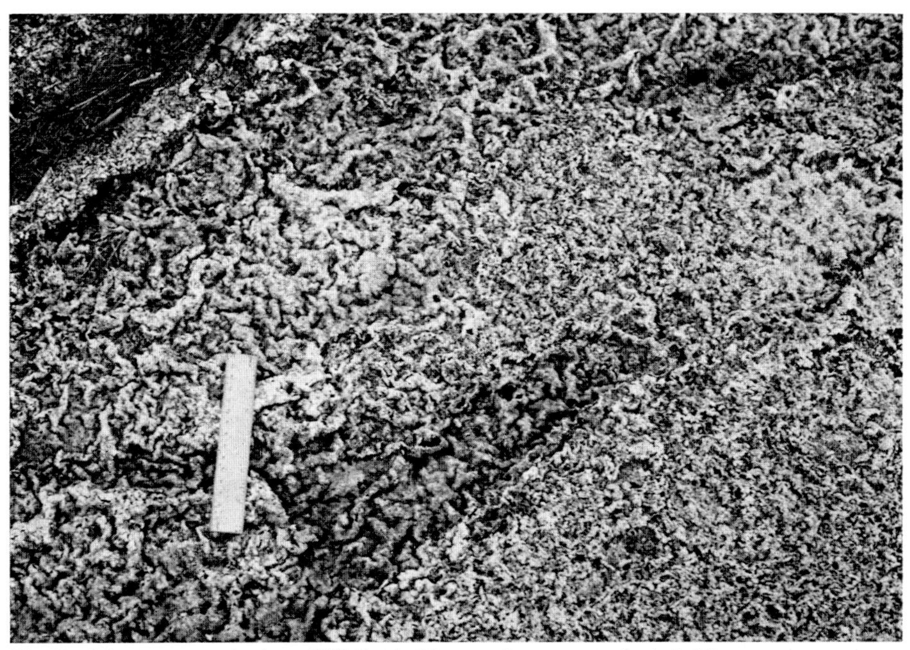

Fig.89. Salt crust on point bar, Cliff Creek. Note surface texture. Scale is 15 cm.

Fig.90. Salt ridges on floodplain, Cliff Creek. Ridges are co-extensive with salt crust and result from expansion as the crust accumulates additional evaporites. Scale is 15 cm.

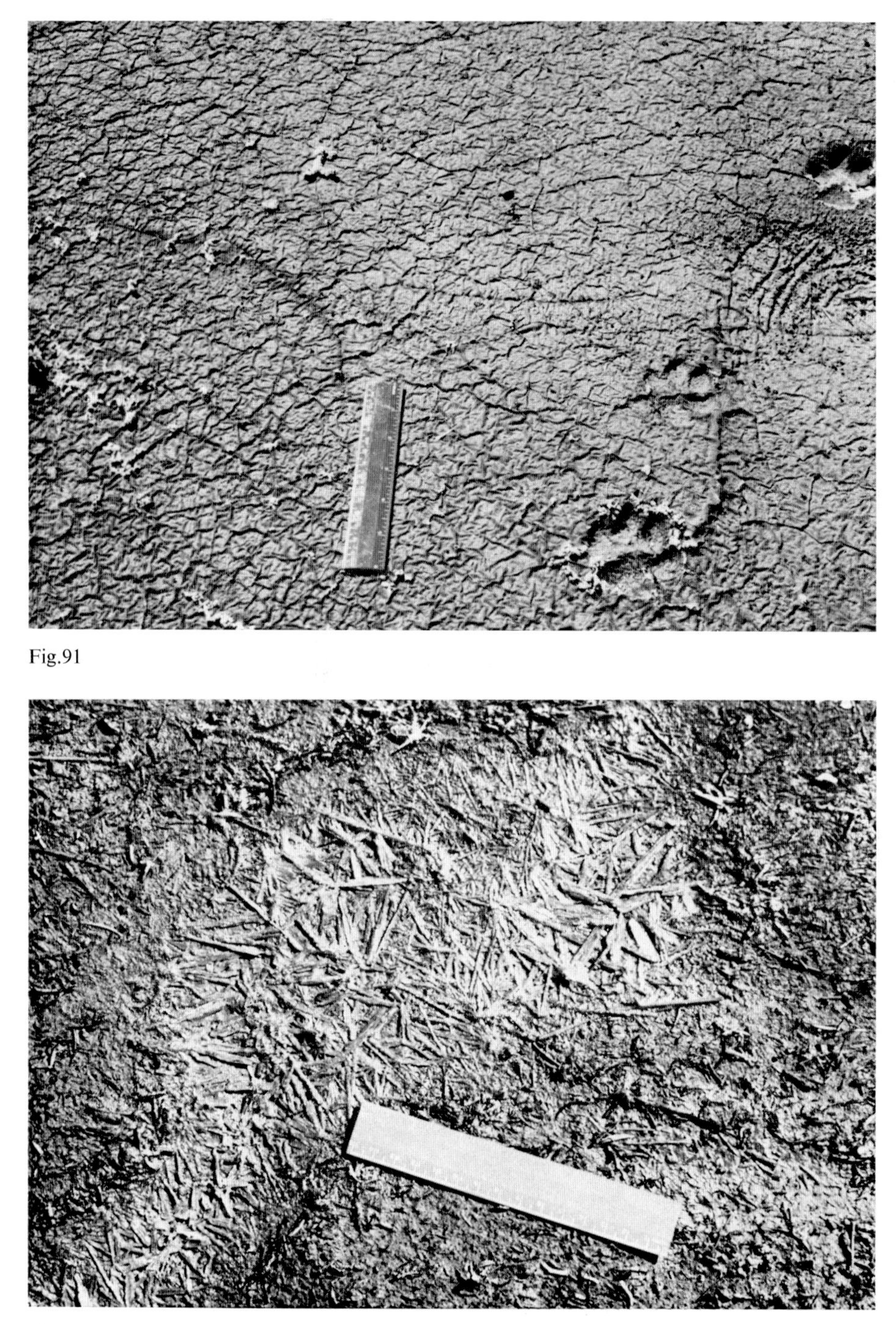

Fig.91

Fig.92

Crystal molds are a final saline structure. Relatively large bladed crystals of gypsum were observed growing in scattered depressions along the channel floor at Cliff Creek (Fig.92).

## Occurrence

Saline features were found along only a few streams. Salt crusts and ridges are

Salt crust

| | | | | |
|---|---|---|---|---|
| – – – – – – – – – – – – – | – – – – | – – – – | c a – t a | – – c – |

Salt ridge

| | | | | |
|---|---|---|---|---|
| – – – – – – – – – – – – – | – – – – | – – – – | – a – – a | – – c – |

Wrinkled surface

| | | | | |
|---|---|---|---|---|
| – – – – – – – – – – – – – | – – – – | – – – – | c c – – – | – – r – |

Crystal mold

| | | | | |
|---|---|---|---|---|
| – – – – – – – – – – – – – | – – – – | – – – – | – t – – – | – – – – |

both extensively developed and, where they occur, are relatively abundant. These features are found on low floodplains and bars, rather than in channels. Wrinkled surfaces are present in channels. Crystals are locally abundant in small depressions that hold ponded water.

## Origin

Salts crusts form from the evaporation of interstitial water. Saline minerals are leached from soils in the drainage basin. High interstitial fluid concentrations may

Fig.91. Wrinkled mud film on channel floor, Cliff Creek. Locally this surface has dried and polygonal shrinkage cracks have developed. On adjacent sand bars, the structure is replaced by reverse curls, with salt crusts on the tops of bars. Note the salt encrustation along the margins of coyote tracks where the mud has been squeezed up above the surrounding level. Length of scale is 15 cm.

Fig.92. Gypsum crystals growing in depression on point bar, Cliff Creek. Other nearby depressions still contained water. Scale is 15 cm.

also result from the solution of mineral fragments, such as gypsum, which survive transportation and are buried in the alluvium. Brackish or mineral-rich spring water also adds to the mineral content. Whatever the source, surface evaporation draws water upward and the dissolved solids are precipitated as a crust on low bars and floodplains.

As salt continues to accumulate the crust expands. The first expansion feature that develops is slight doming. With continued expansion, the crust fails and buckles upward along fractures, forming a system of connected ridges. Further growth accentuates the ridges.

Similarly, the wrinkled mud film may also be an expansion feature, although the exact mechanism is not known and no salt crystals were observed. Alternatively, the wrinkled film may record a volume decrease in the sand substrate without a corresponding decrease in the surficial mud layer. Possible causes are the withdrawal of interstitial water by capillary action to topographically higher areas where evaporation is rapid, or compaction and settling of the water saturated sand. The former explanation agrees with the postulated origin of the salt crusts.

### Preservation and occurrence in sedimentary rocks

The preservation of thin evaporite beds in a dominantly fresh water environment is unlikely. If the evaporites escape solution by rain or stream water, they will probably be dissolved by ground water after burial. However, some salt crusts do survive, as shown by exhumed evaporite layers now exposed in arroyo walls. Surface features such as domes and ridges are probably not preserved or are lost during remobilization after burial. The preservation of crystal cavities is more likely and this structure is present in some abundance in sedimentary rocks. The preservation of wrinkled mud films is also probable, following burial. However, we do not know of any reports of this structure in sedimentary rocks. Possibly, some reports of textured surfaces are in reality wrinkled mud films.

### Other occurrences

Evaporite beds and crystal cavities are common in lacustrine, tidal and restricted marine sediments. In contrast, saline features are rare along streams. In fluvial rocks, most saline features are probably post-depositional in origin.

### Significance

The saline features are not indicative of fluvial environments. If found in fluvial rocks, they are indicative of arid climates, saline minerals and, possibly, ephemeral streams.

## Selected reference

Kendall, C. G. St. C. and Skipwith, P. A. d'E., 1969. Holocene shallow-water carbonate and evaporite sediments of Khoral Bazam, Abu Dhabi, southwest Persian Gulf. *Bull. Am. Assoc. Petrol. Geologists*, 53: 841–869.

# TRACKS, TRAILS, BURROWS
(Lebensspuren, ichnofossil, trace fossil)

## Description

Tracks and trails encompass a variety of organic surface markings. Along the ephemeral streams studied, the most common tracks included coyote, cow, horse and other mammal footprints (Fig.93). Bird tracks are locally abundant, especially around waterholes (Fig.94). In addition to vertebrate footprints, invertebrate trails also are present. Two types of feeding trails were identified. The most common is a long, narrow, shallow, surface trail arranged in broad curves. The

Fig.93. Footprints on channel floor, Dripping Rock Creek. Large tracks were made by a coyote. Also visible are the tracks of a large bird and of a smaller mammal. Raindrop impressions are abundant. Pen is 13.5 cm long.

Fig.94

Fig.95

second pattern is a shallow burrow with a crude dendritic geometry (Fig.95).

**Occurrence**

Tracks and trails are ubiquitous features along ephemeral streams. Although they constitute relatively minor structures at most localities, several tracks or trails can always be found without undue search.

Track, trail

| art – rtr – r – rr | rtrr | cctr | crrcr | crcr |
|---|---|---|---|---|

Tracks and trails, as "accidential" structures, are not related to other structures and are found in all subenvironments.

**Origin**

The origin of vertebrate foot prints is self explanatory. These markings are accidental in that they are made by an organism that is outside of the system. That is, properties of the sediment do not directly influence the occurrence of the structure (with the partial exception of sediment consistency). Thus, the structure is independent of sediment and the environment of deposition.

The origin of markings interpreted to be invertebrate trails is more problematical. The study of trace fossils is currently an area of active research, and the emphasis is placed not on taxonomy but on environmental interpretation. Unlike vertebrate footprints, which permit some reconstruction of the animal, invertebrate trails and burrows usually reveal little about the animal's morphology or identity. Rather, trails record a behavior pattern, often food gathering. Inasmuch as behavior is related to environment, trails and burrows are useful in environmental reconstruction (Seilacher, 1964, pp.296–316).

**Preservation and occurrence in sedimentary rocks**

Vertebrate footprints are rare fossils. The preservation of this surface structure

---

Fig.94. Bird tracks on channel-fill mud at mouth of East Steinaker Draw. The extensive tracks on this surface developed less than 24 h after deposition of the mud.

Fig.95. Invertebrate burrows on channel bank, Stinking Water Creek. Large pattern in center is similar to *Chondrites*. Thin, curved burrows to left resemble *Gordia*. The origin of the rectangular structure in the background is uncertain but it may be micro-slumping. Sparse raindrop impressions are visible. Pen is 13.5 cm long.

requires rapid burial. However, in contrast to less common surface markings (tool marks, grooves, and so forth) the large numbers of footprints makes preservation of some likely as casts or molds.

The occurrence of invertebrate trails and markings is more favorable for preservation. Surface trails (pascichnia) are less likely to survive than are burrows (fondinichnia) and are relatively common in some sedimentary rocks.

### Other occurrences

Footprints are present in all terrestrial and paralic environments that are not covered by more than a few feet of water. Invertebrate trails are less common in terrestrial settings than they are in marine environments, where they are abundant. The long, curved surface trail reported here resembles *Gordia* and the dentritic burrow has some similarity to *Chondrites* (Häntzschel, 1962, pp.W187–W188, W194).

### Significance

These markings are not diagnostic of fluvial environments. Footprints are indicative of terrestrial environments and tidal flats. Trails and burrows are rare in pre-Cretaceous fluvial beds but are much more abundant in later fluvial sequences.

### Selected references

Häntzschel, W., 1962. Trace fossils and problematica, In: R. C. Moore (Editor), *Treatise on Invertebrate Paleontology, part W.–Geol. Soc. Am.*, pp.W177–W245.

Seilacher, A., 1964. Biogenic sedimentary structures. In: *Approaches to Paleoecology*. Wiley, New York, N.Y., pp.296–316.

## RAINDROP IMPRESSION

### Description

Raindrop impressions are small, hemispherical depressions in the surface of fine grained sediment. Many of them have rims that project above the surrounding surface (Fig.96). Diameter of the depression varies, reaching a maximum of 1.5 cm. Raindrop impressions are developed in sets and individual depressions may overlap.

### Occurrence

Raindrop impressions are widespread structures that form along most streams.

Raindrop impression

| c c t r r t – t c – – r | c t c r | r r t – | a c r c r | r r c t |
|---|---|---|---|---|

Because the structure forms most easily in mud, it tends to be most abundant along the lower reaches of streams. Along upper reaches, raindrop impressions are found most frequently in small depressions, rather than over large areas. The structure is most common in channels rather than on bars.

## Origin

Raindrop impressions are impact pits left by rain falling on sediment (especially mud). Proper consistency of the mud is important because too soft mud will not support the pit walls and too hard mud will not crater. Thus, the structure forms soon after exposure of the mud before it has dried and developed a hard surface rind.

## Preservation and occurrence in sedimentary rocks

As with other surface markings, preservation of raindrop impressions is unlikely.

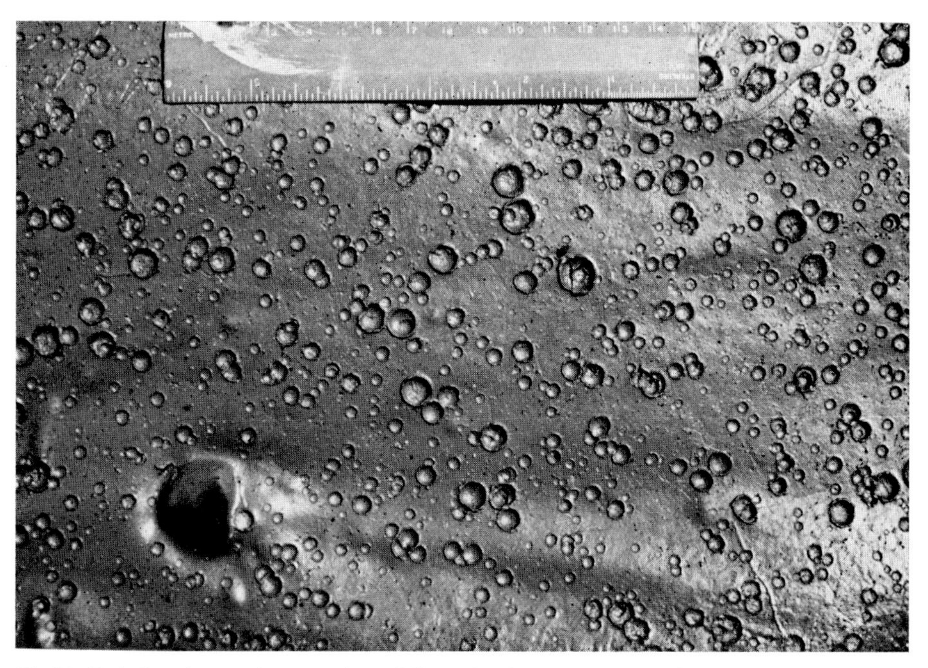

Fig.96. Raindrop impressions on channel floor, Sand Wash. Length of scale is 15 cm.

Adding to the difficulty of preservation is the preferred occurrence in channels, where subsequent floods will scour. Therefore, the structure is rare in sedimentary rocks but occurrences have been found (see references in Shrock, 1948, p.141).

### Other occurrences

Raindrop impressions can form in any environment where soft mud is exposed to the atmosphere. Floodplains, lake margins and tidal flats are excellent settings for raindrop impressions.

### Significance

Raindrop impressions are not diagnostic of fluvial environments. In channel deposits they are indicative of ephemeral flow.

### Selected references

Lyell, C., 1851. On fossil rain marks of the Recent, Triassic and Carboniferous periods. *Q. J. Geol. Soc. Lond.*, 7: 238–247.

Twenhofel, W. H., 1921. Impressions made by bubbles, rain-drops, and other agencies. *Bull. Geol. Soc. Am.*, 32: 359–372.

## TEXTURED SURFACE
(eolian microridges, wrinkle marks)

### Description

Textured surfaces are characterized by small, discontinuous, more or less parallel ridges and swales. Orderliness of the pattern varies from poor (Fig.97) to good (Fig.98), giving the appearance of small, linear ripple marks.

### Occurrence

Textured surfaces occur only on well-sorted dry sand. The structure is widespread

Textured surface

| c r c c c c a c c r r c | t t c c | a c t r | r r r a r | – – – t |
|---|---|---|---|---|

and is developed along most streams. Textured surfaces are most frequently formed on point and longitudinal bars and on floodplains. Occurrences of the structure in channels are rare.

## Origin

Textured surfaces were observed to form by rain beating down on loose sand. The combination of impact on the surface and sheetflow down the slope produces short, discontinuous ridges subparallel with the strike. Where sheetflow is dominant over impact, the ridge pattern is regular. In contrast, where impact is more significant, the degree of parallelism is poor. Loose sand is essential; finer grain sizes are too cohesive as is muddy sand. Along ephemeral streams such sand is most common on point bars and on floodplains. Sands in the channel are either muddy or protected by a mud film.

## Preservation and occurrence in sedimentary rocks

The occurrence of textured surfaces only on loose sands limits the possibility of preservation. However, exposure to the atmosphere produces rather than destroys the structure; so rapid burial is not so important as with most surface markings. Therefore, we conclude that the structure can be preserved and should be present in sedimentary rocks. We have noted similar markings in a number of formations.

Fig.97. Textured surface on point bar, Cliff Creek. Sediment is loose sand. This is an example of a disordered pattern. In other places the ridges occur in a well-defined subparallel trend. Fifteen-centimeter ruler for scale.

Fig.98. Fairly well-ordered textured surface on lateral bar, Halfway Hollow. Length of pen is 13.5 cm.

## Other occurrences

Similar markings have been described from beach sand and called eolian microridges (Hunter, 1969). Other examples have been reported from tidal flats and related environments (Teichert, 1970). Some similar structures have been called interference ripple marks (McKee, 1954, pp.60,61) or eroded symmetrical ripple marks (Shrock, 1948, p.119).

## Significance

Textured surfaces are not diagnostic of fluvial environments. If found in fluvial rocks they are indicative of high bars or floodplains and of exposure.

## Selected references

Hunter, R. E., 1969. Eolian microridges on modern beaches and a possible ancient example. *J. Sediment. Petrol.*, 39: 1573–1578.

Teichert, C., 1970. Runzelmarken (wrinkle marks). *J. Sediment. Petrol.*, 40: 1056–1057.

Twenhofel, W. H., 1921. Impressions made by bubbles, rain-drops and other agencies. *Bull. Geol. Soc. Am.*, 32: 359–372.

ALGAL MAT
(stromatolite)

ALGAL MOUND
(stromatolite, algal ball, algal biscuit)

## Description

Algal structures occur in two forms; thin mats and small mounds. Mats are less than 0.2 cm thick and cover 3–4 $m^2$ (Fig.99). Algal mounds are a few centimeters to about 0.5 m in diameter, 3–10 cm in height, and have a cauliform or botryoidal upper surface (Fig.100). The bottom surface of mounds is generally concave downward. Larger mounds are laminated; small mounds and mats are composed of single layers.

## Occurrence

Algal mounds are widespread and common on floodplains and uplands. Algal mounds were not observed in stream channels or on bar surfaces.

Algal mat

| | | | | |
|---|---|---|---|---|
| - - - - - - - - - - - - - | t - - - | - - - - | - t - - - | t - - - |

Algal mound

| | | | | |
|---|---|---|---|---|
| - c c c c c a - c r c r | - - c r | - t r t | c c a c c | c c c c |

In contrast, algal mats are a rare structure restricted to channel scours.

## Origin

Both algal mounds and mats are biogenic structures formed by algal filaments. Algal mats grow on the bottoms of flooded depressions. Algal mounds grow on high surfaces that are wetted only rarely. Rapid growth of algae probably occurs immediately after a rainstorm and continues until dehydration stops growth. Algal mounds are most abundant around the bases of sage brush plants where the partial shade keeps the surface wet for a longer period.

## Preservation and occurrence in sedimentary rocks

The preservation of non-calcareous algae is unlikely and has occurred in the past

Fig.99. Algal mats floating on ponded water in scour hole, Cliff Creek. Maximum length of algal mats is about 25 cm.

only rarely. Under anaerobic conditions, thin carbon films or masses might survive.

## Other occurrences

Algal mats and stromatolites are uncommon along streams but are widespread in lacustrine, paralic and shallow-marine environments (Bradley, 1929; Carozzi, 1962; Choquette and Traut, 1963; Wolf, 1965; Kendall and Skipwith, 1969).

## Significance

Algal mats and mounds are not diagnostic of fluvial environments. If found in fluvial rocks they may be indicative of channel (mats) conditions or floodplains (mounds). Mounds are indicative of exposure.

## Selected references

Bradley, W. H., 1929. Algae reefs and oolites of the Green River Formation. *U.S. Geol. Surv. Profess. Pap.*, 154: 203–223.

Fig.100. Algal mounds on floodplain, Coyote Wash. Fifteen-centimeter ruler for scale.

Carozzi, A. V., 1962. Observations on algal bioherms in the Great Salt Lake, Utah. *J. Geol.*, 70: 246–253.

Choquette, P. W. and Traut, J. D., 1963. Pennsylvanian carbonate reservoirs, Ismay Field, Utah and Colorado. In: *Shelf Carbonates of the Paradox Basin–Four Corners Geol. Soc., Field Conf., 4th*, pp.157–184.

Kendall, C. G. St. C. and Skipwith, P. A. d'E., 1969. Holocene shallow-water carbonate and evaporite sediments of Khoral Bazam, Abu Dhabi, southwest Persian Gulf. *Bull. Am. Assoc. Petrol. Geologists*, 53: 841–869.

Wolf, K. H., 1965. Petrogenesis and paleoenvironment of Devonian algal limestones of New South Wales. *Sedimentology*, 4: 113–178.

# GAS BUBBLE

(gas pit, bubble impression, pit-and-mound structure)

## Description

Gas bubbles are small, low relief, hemispherical, hollow domes that occur in patches along the stream bed (Fig.101). Within each patch there are a large number of individual domes, almost completely covering the surface. Isolated or sparse bubbles are less common.

### Occurrence

Gas bubbles occur in fine sands on channel floors; none were observed on bars or

Gas bubble

| | | | | |
|---|---|---|---|---|
| – – – – – – – – – – – – – | t – t – | – – – – | – c – r – | – – r – |

on floodplains. The structure is common only at one locality, Cow Wash. Along other streams, it is a minor structure if present.

### Origin

Gas bubbles are a surface expression of sediment compaction. Rapidly deposited channel sand contains large amounts of trapped water and air. During compaction the water and air are expelled from the sediment, together with any gas generated by decaying organic matter. At the surface the gas forms bubbles trapped beneath an impermeable surface film. Pits, which may form if the bubble breaks the surface, were not observed along these streams.

### Preservation and occurrence in sedimentary rocks

Preservation of small surface bubbles is unlikely. Even if preserved, the recognition of the structure after compaction and lithification would be difficult. Accordingly, the extreme scarcity of gas pits and bubbles in sedimentary rocks is understandable. In contrast, the movement within the sediment that causes gas bubbles leads to structures that can be readily preserved as disturbed bedding.

### Other occurrences

Gas bubbles and pits are developed in fluvial, swamp, spring, bog, tidal and deltaic settings. Probably they also are found in deep sea settings.

### Significance

Gas bubbles are not diagnostic of fluvial environments. They are indicative of rapid, episodic deposition.

### Selected references

Cloud, P. E., Jr., 1960. Gas as a sedimentary and diagenetic agent. *Am. J. Sci.*, 258-A: 35–45.
Emery, K. O., 1945. Entrapment of air in beach sand. *J. Sediment. Petrol.*, 15: 39–49.

Fig.101. Gas bubbles on braided channel bottom, Cliff Creek. Each bubble is hollow beneath the thin (less than 1/5 cm) surface film. Length of scale is 15 cm.

Kindle, E. M., 1916. Small pit and mound structures developed during sedimentation. *Geol. Mag.*, 3: 542–547.

Maxon, J. H., 1940. Gas pits in non-marine sediments. *J. Sediment. Petrol.*, 10: 142–145.

Shrock, R. R., 1948. *Sequence in Layered Rocks*. McGraw-Hill, New York, N.Y., 507 pp.

## SAND VOLCANO
(patterned cones)

### Description

Sand volcanoes are small conical mounds of sand with a smaller conical depression

Fig.102. Sand volcanoes in quicksand filling channel, at confluence of Twelvemile Wash with Green River. Gas pits and raindrop impressions are also abundant. Length of twig in foreground is about 12 cm.

at the apex (Fig.102). Size ranges from less than 2.5 cm in diameter up to 5 cm. Heights rarely exceed 3 cm.

### Occurrence

Sand volcanoes are rare sedimentary structures that develop in quicksand. The three occurrences reported here were in channel fills, but inasmuch as quicksand also occurs on longitudinal and transverse bars, the structure is probably not

Sand volcano

| t - - - - - - - - - - - - | t - - - | - - - - | - t - - - | - - - - |
|---|---|---|---|---|

restricted to the channel floor. Because the structure is so rare, generalizations concerning association or position along channels are not possible.

### Origin

Sand volcanoes are a surface expression of sediment compaction and dewatering.

As the sediment settles, large amounts of interstitial water are expelled. If the water migrates upward along definite zones, small springs may form at the surface. Sand and mud particles carried upward by the water are deposited in a cone about the spring. The relative size of the cone is a function of spring activity and duration.

## Preservation and occurrences in sedimentary rocks

Sand and mud volcanoes are rare sedimentary structures. As with gas bubbles, preservation of the surface feature is unlikely, although the resulting disturbed bedding is more easily preserved. Patterned cones (Boyd and Ore, 1963) are probably related structures.

## Other occurrences

Mud volcanoes up to a meter in height are relatively common in hydrothermal areas. Small sand volcanoes resemble closely the mounds that surround some invertebrate burrows in tidal and shallow marine settings.

## Significance

Sand volcanoes are not diagnostic of fluvial environments but are indicative of rapid, episodic deposition.

## Selected references

Allen, J. R. L., 1961. Sandstone-plugged pipes in the Lower Old Red Sandstone of Shropshire, England. *J. Sediment. Petrol.*, 31: 325–335.

Boyd, D. W. and Ore, H. T., 1963. Patterned cones in Permo-Triassic red beds of Wyoming and adjacent areas. *J. Sediment. Petrol.*, 33: 438–451.

Gill, W. D. and Kuenen, P. H., 1958. Sand volcanoes or slumps on the Carboniferous of County Clare, Ireland. *Q. J. Geol. Soc. Lond.*, 113: 441–460.

*Chapter 5*

# STRATIFICATION

Stratification is more difficult to classify than sedimentary structures. Sedimentary structures are relatively discrete features that do not grade continuously from one form to the next. In contrast, stratification types are less distinct and the nomenclature of stratification is chaotic, as exemplified by the terminology for cross-stratification (Potter and Pettijohn, 1963, pp.68–75). The classification of strati-

TABLE XII

Classification of stratification types

| | | | |
|---|---|---|---|
| Bedding type | planar | horizontal parallel | |
| | | horizontal discontinuous | |
| | | lenticular | |
| | | graded | |
| | cross-stratified | micro cross-stratification | |
| | | festoon | |
| | | ripple | Type-A |
| | | | Type-B |
| | | | sinusoidal |
| | | inclined | scour fill |
| | | | channel fill |
| | | low-angle | |
| | | avalanche-front | |
| | | backset | |
| | disturbed | slump | convolute |
| | | | flame structure |
| | | | flow roll |
| | | bioturbation | burrowed |
| | | | root-disturbed |
| | | desiccation | mud-pebble |
| | | | desiccation crack |
| | | compaction | |
| | | erosion | stepped erosion surface |

fication types used here (Table XII) is for convenience only and is not intended as a general classification.

An understanding of stratification is essential to the interpretation of fluvial rocks. In most fluvial units sedimentary structures are rare but planar bedding and cross-stratification are abundant. In this chapter we describe and illustrate the variety of stratification types that exist along ephemeral streams. In general, the nomenclature of McKee and Weir (1953, pp.382–384) is used. In the discussion, the terms listed below are defined as follows:

*Stratification:* a general term that describes the property of layering in sedimentary rocks.
*Stratum:* basic unit of stratification consisting of each individual layer; smallest subdivision possible.
*Bed:* stratum more than 1 cm in thickness.
*Lamina:* stratum less than 1 cm in thickness.
*Set:* group of strata that are essentially conformable and continuous.
*Sedimentation unit:* one or more sets deposited continuously under uniform or continuously varying condition; represents a single episode of sedimentation. (For a general discussion of sedimentation units, the reader is referred to Jopling, 1964).

## HORIZONTAL PARALLEL STRATIFICATION
(uniform stratification)

### Description

Horizontal parallel stratification consists of thin, even strata (Fig.103). Each stratum is parallel with the others and the entire set is approximately horizontal. Individual strata are laterally continuous for distances exceeding 30 cm. Strata are up to 0.5 cm thick and occur in sets 5–15 cm thick. The overall appearance is of fine, almost "perfect" layers.

### Occurrence and associated sedimentary structures

Horizontal parallel stratification is found in thin sets in both bar and channel deposits. Although this type of bedding immediately underlies the surface at many localities, it is volumetrically minor. In most trenches horizontal parallel stratification accounted for only a few percent of the deposit. Horizontal parallel stratification is present in most point bars and is common in the downstream segments of channels. In contrast, this type of bedding is rare upstream.

Horizontal parallel stratification is associated with parting lineation, which is the splitting property of this type of bedding. Therefore, the relative abundance of parting lineation can be used to estimate the abundance of horizontal parallel stratification.

Horizontal parallel stratification is found in sequences with other bedding types (Fig.104). Commonly, horizontal parallel stratification is present at the top of a sedimentation unit and micro cross-stratification occurs immediately below the horizontal stratification. Locally, thin mud films cover sedimentation units.

## Origin

The clear association of horizontal parallel stratification and parting lineation requires that the two features have a common origin. Formation by low velocity currents is indicated by the sequence of bedding types. Within a sedimentation unit, horizontal parallel stratification is found above micro cross-stratification and under a mud film. Thus, the bedding results from currents that are intermediate between those strong enough to cause saltation (cuspate ripples and micro cross-stratification) and those so weak that the suspended load (mud film) is

Fig.103. Horizontal parallel stratification in longitudinal bar, Twelvemile Wash. Flow is from left to right. Thickness of the laminae is greater than normal. Length of scale is 15 cm.

Fig.104. Horizontal parallel stratification exposed in pit in channel, Twelvemile Wash. Horizontal parallel stratification is developed in a thin set at the top above the micro cross-lamination. Horizontal discontinuous stratification is exposed below the micro cross-lamination. Current is from right to left. Surface is marked by parting lineation. This pit shows the approximate abundance of horizontal parallel stratification in comparison with other bedding types.

deposited. Despite the relatively low flow velocities indicated for horizontal parallel stratification, upper-flow regime conditions are probable, resulting from the relatively fine grain size in comparison with micro cross-stratification, and (or) shallower depths as the flood wanes and the bottom is elevated. The latter factor is probably the more significant in this study because grain-size differences between the stratification types within sedimentation units are minor.

Despite the strong evidence from flume experiments, lower-flow regime conditions cannot be ruled out. Sorby's pioneering observations show that transportation does occur at velocities less than those required to initiate saltation and rippling, resulting in plane beds. Although this result has not been repeated by more recent experiments, this may just reflect an oversight because of a general neglect of low flow velocities. It is also possible that horizontal parallel stratification may be analogous to ripple stratification, representing deposition from overloaded currents. Under this condition, plane stratification may result from currents just able to cause transportation. Both of these possibilities deserve further experimental study.

### Preservation and occurrence in sedimentary rocks

The preservation of horizontal parallel stratification is not as likely as the preservation of most other types of bedding. Bedding that occurs at the surface is subject to destruction by wind ablation, mud cracking, animals or subsequent floods. The last cause is probably the most significant. Each new flood scours the deposits of previous floods and deposits a new unit. Repeated floods result in nested, partially eroded sedimentation units, each of which is missing the upper part of the sequence. Thus, the same reasons that account for the scarcity of bedding-plane structures in fluvial deposits also explain the scarcity of horizontal parallel stratification.

### Other occurrences

Horizontal parallel stratification associated with parting lineation occurs in beds formed in several environments. These structures are common in fluvial rocks, especially in point-bar deposits, and in deposits of shallow-marine or lacustrine environments. Similar bedding is abundant in cross-stratified eolian sandstone, but the sets are inclined rather than horizontal. Horizontal parallel stratification is also found in turbidites but is rare (E. F. McBride, 1972, personal communication).

### Significance

Horizontal parallel stratification is not diagnostic of fluvial environments. It is formed by low-velocity currents. Flow regime is uncertain. Associated parting lineation indicates the current direction.

### Selected reference

Smith, N. D., 1971a. Pseudo-planar stratification produced by very low amplitude sand waves. *J. Sediment. Petrol.*, 41: 69–73.

## HORIZONTAL DISCONTINUOUS STRATIFICATION
(planar bedding, plane bedding)

### Description

Horizontal discontinuous stratification superficially resembles horizontal parallel stratification (Fig.104). However, close inspection reveals several significant differences. Strata thickness, although still fine, is thicker in horizontal discon-

Fig.105. Horizontal discontinuous stratification in channel deposit, Twelvemile Wash. Current is from left to right. Two sedimentation units are present, separated by an erosional surface in upper third of the photograph. Note stepped profile of erosion surface. Fifteen-centimeter ruler for scale. Lower sedimentation unit consists entirely of horizontal discontinuous stratification. Upper unit shows thick sequences of horizontal discontinuous stratification overlain by micro cross-stratification.

tinuous than in horizontal parallel bedding. Furthermore, the horizontal discontinuous bedding occurs in thick sets, up to a meter thick, in contrast to the thin sets of horizontal parallel bedding. Finally, horizontal discontinuous stratification is characterized by micro-lensing. Individual strata are not laterally continuous. Rather, the strata pinchout over distances of a few meters (Fig.105, 106).

## Occurrences and associated sedimentary structures

Horizontal discontinuous stratification is associated with streaming lineation, which is the surface expression of this bedding type. As with horizontal parallel stratification and parting lineation, splitting of horizontal discontinuous bedded rocks should reveal streaming lineation on the bedding planes.

Horizontal discontinuous stratification is present in most channel deposits and is the most abundant bedding type observed. In bar deposits, this type of

Fig.106. Horizontal discontinuous stratification in channel deposit, Coyote Wash. Flow is from left to right. Note small-scale lensing of strata. Scale is in inches and centimeters. Micro cross-lamination at extreme top of photograph.

bedding is less common and is subordinate to low-angle cross-stratification. In sedimentation units, horizontal discontinuous stratification is usually lowest in the sequence and rests on an erosional surface.

McKee et al. (1967, p.829) estimated that horizontal strata constituted 90–95% of all deposits in flood deposits that they studied at Bijou Creek, Colorado. Their pictures indicate that this was horizontal discontinous stratification for the most part.

## Origin

Horizontal discontinuous stratification is deposited by high-velocity currents at, or just over, the threshold of the upper-flow regime. This origin is indicated by the association of horizontal discontinuous stratification with streaming lineation and by its position within channel-fill sequences. As flood waters wax, the channel bottom is scoured and little deposition occurs. As the flood crest passes, erosion is replaced by deposition during the waning phase. Initially, currents are vigorous and capable of producing upper-flow regime bedforms. As the flood waters continue to fall, the current velocity decreases and a corresponding sequence of bedforms is produced. Thus, the channel-fill deposit is a sequence from high to low velocity and the position of horizontal discontinuous stratification at the base of this sequence indicates formation by high-velocity currents.

### Preservation and occurrence in sedimentary rocks

Because of its occurrence at the base of a sedimentation unit, horizontal discontinuous stratification is in one of the most favorable positions for preservation. This bedding type rarely is present at the surface, but is generally buried by 30 cm or more of sediment. Another factor favoring preservation is the abundance of horizontal discontinuous stratification. In rocks, this type of bedding generally has not been differentiated from horizontal parallel stratification by most workers and both are reported together. It is likely that much of the horizontal lamination in coarse sandstone is horizontal discontinuous stratification.

### Other occurrences

Horizontal discontinuous stratification can form wherever high-velocity currents exist; beaches and some tidal flats are favorable settings.

### Significance

Horizontal discontinuous stratification is indicative of high flow velocity and conditions within the lower part of the upper-flow regime. It is not diagnostic of environment of deposition. In fluvial rocks, horizontal discontinuous stratification is indicative of proximal deposits and channel fills. Current directions can be obtained from associated streaming lination.

### Selected references

Jopling, A. V., 1960. *An Experimental Study of the Mechanics of Bedding*. Thesis, Harvard Univ. 358 pp.

Simons, D. B. and Richardson, E. V., 1961. Forms of bed roughness in alluvial channels. *Proc. Am. Soc. Civil Eng., J. Hydraulics Div.*, 87: 87–105.

## LENTICULAR STRATIFICATION

### Description

The term lenticular stratification describes the relationship between stratification sets. Stratification thickness and type within sets is variable but horizontal and low-angle cross-lamination are dominant. Where gravel is present, stratification within sets is generally obscure. Individual sets commonly are 5–15 cm thick and persist laterally for several meters before pinching out. Most pinchouts apparently are the result of non-deposition, although some of them are the result of subsequent erosion. Commonly, the attitude of lenticular sets is horizontal.

**Occurrence**

Because of its large size, lenticular stratification was only observed in exposures along stream banks (Fig.107,108); trenches are too small to adequately reveal this feature. Consequently, our sample is biased and we can only estimate the distribution of this bedding type. However, judging from the unstable nature of bars and channels in ephemeral streams, we believe that lenticular stratification is probably the major type of relationship between stratification sets.

Fig.107. Lenticular stratification in channel wall, Halfway Hollow. Flow is from right to left. Gravel layers, with crude low-angle cross-stratification and imbrication, were deposited in small channels. Intervening sands, also with low-angle cross-stratification, were deposited as longitudinal bars. Note the size grading in the gravels. Largest boulder (center of photograph) is about 40 cm long.

Fig.108. Lenticular stratification in channel bank, Cliff Creek. Hammer is 30 cm long.

## Origin

Lenticular stratification forms as a result of shifting depositional centers. Given the different subenvironments within a stream and the rapidly shifting regime, it is not surprising that lenses of sediment are deposited atop each other. Unlike sedimentation units, lenticular stratification does not record the steady and progessive change in conditions within a single regime; rather it records fluctuating regimes, as from bar to channel and back again. The result is a composite set of strata that does not share the internal consistency of a sedimentation unit.

## Preservation and occurrence in sedimentary rocks

Lenticular stratification is common in fluvial rocks. Preservation is likely, especially along wide streams where flooding does not scour from valley wall to valley wall.

## Other occurrences

Lenticular stratification is developed in environments characterized by shifting sedimentary regimes. This bedding type is prevalent in braided streams and on some tidal flats.

## Significance

Lenticular stratification is suggestive of fluvial deposition. Its presence is indicative of fluctuating sedimentary regimes.

## GRADED STRATIFICATION
(interrupted graded bedding, fining-upward cycles)

## Description

Graded stratification is characterized by an upward decrease in grain size. Along ephemeral streams, size grading is found at two scales. The largest and most common occurrence is within sedimentation units (see "Point-bar sequence", "Channel-fill sequence") where the sediment of the lower part of the unit is coarser

Fig.109. Graded channel-fill and bar complex, Little Mountain Creek. Gravelly sand of channel-fill grades upward into low-angle cross-stratified sand of point bar. Sedimentation unit is completed by ripple-bedded unit with flame structures and convolute bedding. In the upper fifth of the photo are low-angle cross-strata of a second point bar. Flow is from right to left. Pen is 13.5 cm long.

than that above. Generally, the size range is not great, ranging from medium sand at the base to silt at the top. At this scale, size gradation is present over a distance of 5–15 cm and encompasses several sets of strata (Fig.109).

More typical of traditional graded stratification is an upward decrease in grain size within a single stratum or set. The latter commonly is developed in gravel deposits (Fig.107).

**Occurrence**

Graded sedimentation units are abundant. Every channel or bar sequence that was excavated showed size grading. Graded strata or sets are rare. Only gravel units display an obvious decrease upward in grain size. Other units, however, may have less pronounced grading that would become apparent with detailed size analyses.

**Origin**

Graded bedding in streams results from a decrease in transport velocity. In bar and channel sedimentation units the upward decrease in grain size is accompanied by bedform changes that are indicative of progressively decreasing current velocity. As the flood recedes and velocity slackens, the scoured channel is back-filled with sediment. The first deposits are from still vigorous currents and are relatively coarse. As the velocity continues to decrease, the size of the bed load becomes smaller until, during the last stage, mud settles from water standing in local depressions. Similarly, graded gravel strata also record the passing of a flood. At its height the flood erodes the bottom, leaving gravel as a lag deposit. Subsequently, sand is deposited on the gravel layer as the flood wanes.

**Preservation and occurrence in sedimentary rocks**

Graded sedimentation units commonly are preserved in whole or in part, and constitute a large fraction of alluvium and sedimentary rocks. The preservation of graded strata is less likely, inasmuch as they form within channels and are relatively rare. Nevertheless, graded sets have been observed in fluvial units.

**Other occurrences**

Fining-upward cycles are common in fluvial rocks and are characteristic of point-bar deposits (Visher, 1965; Allen, 1965b,1970a; McCave, 1969; McCormick and Picard, 1969). Similar size trends also are believed to occur in transgressive marine deposits and in parts of regressive marine and deltaic cycles (Visher, 1965). Grading within individual strata also is widespread, occurring in all environments from deep marine through eolian (Shrock, 1948, pp.77,78).

## Significance

Graded stratification is not diagnostic of fluvial environments. However, it is indicative of progressive changes in conditions within a sedimentary regime (in contrast with lenticular stratification).

## Selected references

Allen, J. R. L., 1965b. Fining-upward cycles in alluvial successions: *Geol. J.*, 4: 229–246.
Allen, J. R. L., 1970a. Studies in fluviatile sedimentation: a comparison of fining-upwards cyclothems, with special references to coarse-member composition and interpretation. *J. Sediment. Petrol.*, 40: 293–323.
McCave, I. N., 1969. Correlation of marine and non-marine strata with example from Devonian of New York State. *Bull. Am. Assoc. Petrol. Geologists*, 53: 155–162.
Shrock, R. R., 1948. *Sequence in Layered Rocks.* McGraw-Hill, New York, N.Y., 507 pp.

# MICRO CROSS-STRATIFICATION
(small-scale trough cross-stratification, wispy cross-stratification, ripple-drift cross-stratification)

# FESTOON CROSS-STRATIFICATION
(medium-scale trough cross-stratification)

## Description

Micro cross-stratification is composed of small, scoop-shaped sets of cross-strata. In transverse section, the strata within a set are concave upward and conform to the lower erosional surface. Individual sets are packed together with partial erosion by adjacent and overlying sets (Fig.110–112). In longitudinal view the sets are tabular- or wedge-shaped; strata are slightly concave upward and are inclined downstream. Individual strata and sets are thin. Festoon cross-stratification is a medium-scale (McKee and Weir, 1953) version of micro cross-stratification (Fig.113, 114).

## Occurrence and associated sedimentary structures

Micro cross-stratification and festoon cross-stratification are present in most channel and bar deposits. The bedding commonly develops from cuspate ripple marks, but it can also form from longitudinal ripple marks (Fig.111). Micro cross-

Fig.110

Fig.111

Fig.112. Longitudinal section of micro cross-stratification on point bar, Twelvemile Wash. Flow is from right to left. Low-angle cross-stratification below. This micro cross-stratification is also considered to be ripple-drift stratification, Type-A. Pen is 13.5 cm long.

stratification is found in thin units near the top of bar and channel sequences. Generally, micro cross-stratification is overlain by horizontal parallel stratification and underlain by either horizontal discontinuous (channel) or low-angle cross-stratification (bar) (Fig.112). Because of the thinness of the micro cross-laminated zone, this type of bedding is volumetrically minor. Micro cross-stratification is more common than is festoon cross-stratification. Where both types of bedding occur together, festoon cross-stratification always is found below micro cross-stratification in the same sedimentation unit.

---

Fig.110. Transverse section of micro cross-stratification in channel deposit, Twelvemile Wash. Flow is away from viewer. Horizontal discontinuous stratification below. Note the erosion surface (marked by mud film) above the micro cross-stratification and the small channel to the extreme right. The upper sedimentation unit consists of inclined strata below and horizontal parallel stratification above. Parting-lineation and desiccation cracks are present on the surface. Length of scale is 15 cm.

Fig.111. Micro cross-stratification associated with longitudinal ripples, Coyote Wash. Flow is away from viewer. Festoon cross-stratification horizontal discontinuous stratification below. Scale is in inches and centimeters.

Fig.113. Transverse section of festoon cross-stratification in channel deposit, Coyote Wash. Flow is away from viewer. Horizontal discontinuous stratification below and micro cross-stratification above. Fifteen-centimeter ruler for scale.

## Origin

Micro cross-stratification and festoon cross-stratification are formed by moderate velocity currents in the lower-flow regime (Harms and Fahnestock, 1965, pp.106, 107). Of the two bedding types, festoon cross-stratification probably records higher current velocities that were sufficient to produce dunes, in contrast with the cuspate ripple marks that are associated with micro cross-stratification (Harms, 1969, pp.378, 379).

Fig.114A. (For legend, see p.160.)

## Preservation and occurrence in sedimentary rocks

Micro and festoon cross-stratification are common in fluvial rocks. The likelihood of preservation along ephemeral streams is limited by the thinness of the units and their occurrence near the top of sedimentation units.

## Other occurrences

Micro and festoon cross-stratification are formed in all fluvial and shallow-marine environments. In fluvial rocks these bedding types are more common in braided rivers than in ephemeral streams (Williams and Rust, 1969, p.657; Smith, 1970,

Fig.114A,B. Transverse section of festoon cross-stratification in channel-bar complex, Twelvemile Wash. Note development of multiple sedimentation units and horizontal discontinuous, inclined, and micro cross-stratification. Bar to left, channel to right. Flow is away from viewer. Length of scale is 15 cm.

pp.2999, 3000). In sedimentary rocks, micro cross-stratification commonly is associated with rib- and-furrow structure (Hamblin, 1961).

## Significance

Micro and festoon cross-stratification are not diagnostic of fluvial environments. They form in response to conditions of moderate flow in the lower-flow regime. Forsets are inclined downcurrent.

## Selected references

Allen, J. R. L., 1963. The classification of cross-stratified units with notes on their origin. *Sedimentology*, 2: 95–114.
Hamblin, W. K., 1961. Micro-cross-lamination in Upper Keweenawan sediments of northern Michigan. *J. Sediment. Petrol.*, 31: 390–401.
McKee, E. D., 1965. Experiments on ripple lamination. *Soc. Econ. Paleontologists Mineralogist, Spec. Publ.*, 12: 66–83.
Picard, M. D. and High, L. R., Jr., 1964. Pseudo rib-and-furrow marks in the Chugwater Formation of west-central Wyoming. *Contrib. Geol.*, 3: 27–31.
Williams, G. E., 1968. Formation of large-scale trough cross-stratification in a fluvial environment. *J. Sediment. Petrol.*, 38: 136–140.

# RIPPLE STRATIFICATION
(ripple-drift stratification, climbing ripples, pseudo cross-stratification)

## Description

Ripple stratification is a type of cross-stratification in which the entire ripple form is preserved. In most cross-stratification only the foresets of ripples or dunes remain, but in ripple stratification the crest and backslope also are preserved. Ripple-stratified units are thinly bedded, and laminae are dominant in comparison with other sizes. Within the rippled set, each lamina usually preserves the ripple form. Ripples are superposed on those below and the form is repeated upward throughout the set. The upward migration of ripple form throughout the set can be vertical, inclined downstream (Fig.115), or inclined upstream (Fig.116).

The ripple stratification described here is the type-B ripple-drift cross-stratification of Jopling and Walker (1968). Characteristics of this type are: continuous laminae across the ripple form, an asymmetrical ripple profile, and migration of the ripple form throughout the set. Type-A ripple-drift cross-stratification corresponds to the micro cross-lamination shown in Fig.112. Because of the poor preservation of ripple form we prefer the name micro cross-stratification. Only isolated examples of sinusoidal ripple stratification were observed.

## Occurrence and associated sedimentary structures

Ripple stratification is a minor bedding type along ephemeral streams. It occurs less frequently than horizontal parallel or micro cross-stratification and accounts for less than a few percent of the bedding types found in the deposits. In channel deposits ripple stratification is restricted to the top portions of bar and channel sedimentation units. These ripple sets are thin, rarely exceeding a few centimeters. Thick sets of ripple stratification are common in high bars or floodplain deposits.

Fig.115

Fig.116

Ripple stratification is associated with a wide variety of ripple marks, cuspate and longitudinal in channel deposits, and linear asymmetric and secondary ripples on bars. Ripple stratification frequently grades laterally into micro cross-lamination or convolute bedding with flame structures.

## Origin

Ripple stratification generally is considered to result from rapid deposition in areas of high sediment input (McKee, 1965, pp.74, 75). As stressed by Jopling and Walker (1968, pp.982, 983) and Allen (1970b, p.20), this structure allows only an estimate of the ratio between the sediment deposition from suspension and the bedload transportation rate. Absolute values of either parameter cannot be determined.

The trend from type-A (here called micro cross-stratification) to type-B (here called ripple stratification) to sinusoidal ripple stratification has been interpreted to reflect: (*a*) increasing ratios of suspended to traction load (Jopling and Walker, 1968, pp.982, 983); (*b*) proximal to distal changes in currents (Allen, 1970b, pp.22, 23); and (*c*) temporal changes with decreasing current strength (Allen, 1970b, pp.23, 24). These interpretations could not be confirmed in this study because of the scarcity of ripple stratification.

## Preservation and occurrence in sedimentary rocks

Ripple stratification occurs widely in sedimentary rocks. Although a relatively minor structure, it is by no means rare in rocks that contain abundant ripple marks. Preservation of the structure along ephemeral streams is limited by the thinness of the units and their occurrence near the tops of sedimentation units. However, preservation is more likely where ripple stratification is present in thick sets.

---

Fig.115. Ripple stratification on point bar. Twelvemile Wash. Channel to right. Flow in channel is away from viewer. Swash over bar is from right to left. Linear asymmetric ripples on bar surface. Ripples are asymmetric and laminae are continuous across ripple. Festoon cross-stratification below. Pen is 13.5 cm long.

Fig.116. Ripple stratification in channel fill from near the mouth of Twelvemile Wash. Flow in channel is from left to right. Ripples developed below the head of the hammer are asymmetric and migrate upstream, possibly reflecting flooding from nearby Green River to form a temporary "estuary". Flow reversal occurred at position shown by the head of the hammer. Isolated zones of micro cross-stratification (type-A ripple-drift stratification) are present. Note the mud clast at the lower right. Hammer handle is 28 cm long.

### Other occurrences

We have observed ripple stratification in shallow lacustrine (Evacuation Creek Member of the Green River Formation) and shallow-marine (Red Peak Member of the Chugwater Group) units. The structure is also characteristic of kame deposits. Ripple stratification, however, is most abundant in fluvial deposits, including floodplains, channels and bars (McKee, 1964, p.280). In flood deposits of the Colorado River of the Grand Canyon, ripple stratification was the most abundant structure (McKee, 1938, p.80).

### Significance

Ripple stratification is suggestive of fluvial deposition, but other environments can not be excluded. Its presence indicates a relatively high rate of deposition in comparison with transportation. Ripple stratification is used for interpreting current velocity and proximal/distal position. Foresets are inclined downcurrent.

### Selected references

Allen, J. R. L., 1970b. A quantitative model of climbing ripples and their cross-laminated deposits. *Sedimentology*, 14: 4–26.
Jopling, A. V. and Walker, R. G., 1968. Morphology and origin of ripple-drift cross-lamination, with examples from the Pleistocene of Massachusetts. *J. Sediment. Petrol.*, 38: 971–984.
McKee, E. D., 1965. Experiments on ripple lamination. *Soc. Econ. Paleontologists Mineralogists, Spec. Publ.*, 12: 66–83.
McKee, E. D., 1966. Significance of climbing-ripple structure. *U.S. Geol. Surv., Profess. Papers*, 550-D: 94–103.
Walker, R. G., 1963. Distinctive types of ripple-drift cross-lamination. *Sedimentology*, 2: 173–188.
Walker, R. G., 1969. Geometrical analysis of ripple-drift cross-lamination. *Can. J. Earth Sci.*, 6: 383–391.

## INCLINED STRATIFICATION
(parallel inclined stratification, scour-and-fill, washouts)

### Description

Inclined stratification consists of a single set of strata that covers an irregular erosional surface. Where the strata fill local depressions, they dip into the hole with a convex-up profile and terminate against the bottom or opposite side. Size ranges from small scours measuring 5–15 cm (Fig.117) to large channels measuring a meter or more (Fig.118, 119). Stratification within each inclined set is thin and is characteristically horizontal discontinuous in type. The inclination generally is

Fig.117. Inclined stratification in small scour in channel fill, Twelvemile Wash. Flow is from left to right. Scour is at the base of the uppermost of two sedimentation units. Notice slight upstream inclination of strata in lower inch of upper sedimentation unit. Horizontal discontinuous stratification is also present. Scale is 15 cm long.

apparent in only one section, transverse for channel fills and longitudinal for scours. Sections at right angles show only horizontal discontinuous stratification or, for small scours, festoon cross-stratification.

**Occurrence**

Inclined stratification is one of the major bedding types in ephemeral stream deposits. Small scours are rare but inclined strata filling channels are abundant. The most common setting for this bedding type is the transition from lateral or point bars into channels. Strata of the bars, usually low-angle cross-strata, can be traced continuously into channel depressions. At bar edges the strata drape over the abrupt slope and dip into channels. Thus, the direction of inclination is toward the center of the channel, rather than downcurrent.

**Origin**

Inclined stratification results from the back-filling of scours as flood waters recede.

The depression is usually asymmetric in transverse section, except for small scours, with a gentle side next to a bar and a steep opposite face. In general, deposition is similar to that of point bars which migrate laterally into channels, resulting in cross-stratification at right angles to the channel direction (Williams, 1966). In contrast to point-bar slip faces, inclined stratification is smaller and is more directly influenced by bottom topography. Deposition in the depression is not by slumping off the bar face. If that were the case individual strata would abut against the bottom at the angle of repose. In inclined stratification the laminae are convex-upward as they pass over the edge of the depression and turn up sharply near the bottom as shown in Fig.118, 119. Thus, deposition is by currents moving along the face, carrying sediments in traction. On the outside next to the steep face, deposition is prevented by the deflected streamlines. With the passing of the flood and the subsequent decrease in current velocity, lateral accretion progressively builds the bar face out and fills the channel.

## Preservation and occurrence in sedimentary rocks

Inclined strata frequently are preserved in ephemeral streams. The depressions

Fig.118. Inclined stratification filling channel, Twelvemile Wash. Point bar to left. Flow is away from viewer. Note the sudden coarsening of sediment at the cut face. Traces of several older inclined sets are visible in the lower six inches. Hammer for scale.

Fig.119. Inclined stratification on edge of point bar, Twelvemile Wash. This photograph is immediately to the left of Fig.118. The large inclined set in the center is continuous with the partially eroded set at the top of the hammer in Fig.118. Note that the cut- and slip-banks have reversed between the two episodes. Several older inclined sets are visible in the lower third of the face. Note the discordance of the inclined strata in contrast to the festoon near the top. Scale is 15 cm long.

formed are quickly buried under 60–100 cm of waning flood sediment. While scour by successive floods is probable, only part of the channel fill is usually removed as the channel shifts its position slightly. A common feature in channel trenches is the presence of several partially eroded sets of inclined strata (Fig.119).

### Other occurrences

Inclined stratification is common in point-bar deposits (McGowen and Garner, 1970, p.86). Similar bedding is not reported from braided streams (Williams and Rust, 1969, pp.655, 656; Smith, 1970, pp.2998–3000). Although cut-and-fill structures are present in shallow-marine and beach deposits, extensive inclined stratification probably is restricted to streams and, possibly, tidal channels.

### Significance

Inclined stratification is suggestive of fluvial deposition and meandering or ephemeral streams. It does not give the current direction. On a large scale, inclined stratification indicates continuous point bar migration; on a small scale with abrupt scours it indicates fluctuating water conditions.

### Selected references

McGowen, J. H. and Garner, L. E., 1970. Physiographic features and stratification types of coarse-grained point bars: modern and ancient examples. *Sedimentology*, 14: 77–111.
Williams, G. E., 1966. Planar cross-stratification formed by the lateral migration of shallow streams. *J. Sediment. Petrol.*, 36: 742–746.

## LOW-ANGLE CROSS-STRATIFICATION
(point-bar cross-stratification, tabular cross-stratification, planar cross-stratification)

### Description

Low-angle cross-stratification forms the largest units of stratification that we observed along ephemeral streams. The sets are about 30 cm thick and may be several meters long. The lower surface of each set can either be non-erosional or planar or both. Strata within the set are straight to slightly concave. Where the strata approach the lower boundary of the set they tend to turn up slightly and are concordant. Inclinations of the strata are low, usually less than ten degrees, and maximum dips are downcurrent.

### Occurrence

Low-angle cross-stratification is one of the major bedding types along ephemeral streams equal in abundance to horizontal discontinuous stratification. Low-angle

Fig.120. Low-angle cross-stratification in point bar, Halfway Hollow. Flow is from right to left. Hammer for scale.

cross-strata are restricted to point and longitudinal bars where they are developed at the bases of sedimentation units. They may comprise 50–90% of the bedding in the bars (Fig.120, 121). Low-angle cross-stratification can grade laterally into horizontal discontinuous or inclined stratification in channel deposits or can be truncated by erosional surfaces with steps. Low-angle cross-stratification commonly grades upward into ripple or convolute stratification.

## Origin

Low-angle cross-stratification records the downcurrent migration of point and longitudinal bars. Deposition by low-velocity currents while the bars are still flooded during waning flood stage is indicated by the fine grain size, the long length of thin strata, and the low angles of inclination. The sediment is finer than in adjacent channels and is deposited over the entire top of the submerged bar. Relatively deep water is shown by the lack of ripple stratification, which is indicative of shoal depths. Although the current velocities are certainly less than in the adjacent channel, as indicated by the grain size, conditions approaching the upper-flow regime for the fine sand are attained, resulting in long, parallel strata and the

Fig.121. Low-angle cross-stratification in longitudinal bar near the mouth of Twelvemile Wash. Flow is from left to right. Floodplain mud deposit at base of face (not shown) is covered by a low-angle set dipping to left (below hammer). This lower sedimentation unit probably records flooding of the creek from the Green River several hundred feet downstream (see Fig.116). A current reversal is shown at the head of the hammer and ripple stratification followed by a low-angle set (dipping to right) is developed. The upper sedimentation unit is capped by micro cross-stratification and desiccation cracks in another mud film.

absence of current ripples. As velocity decreases with the passing of the flood, cuspate ripples and micro cross-stratification are formed on the lower parts of the bar (Fig.121). Higher bar surfaces are barely awash and are covered with linear asymmetric ripple marks or parting lineation. Thus, low-angle cross-stratification on bars is the companion of horizontal discontinuous stratification in channels. Low-angle cross-stratification grades into inclined stratification, from which it is distinguished by finer grain size, continuity of individual strata, consistent internal geometry, and low angle of inclination. Both stratification types record the deposition of plane beds on sloping surfaces, but in different settings. Inclined stratification is deposited in the relatively energetic channel margin environment and low-angle cross-stratification is formed by point and longitudinal bar migration. Furthermore, inclined stratification is deposited on an irregular, erosional surface and low-angle cross-stratification is deposited on a depositional surface of its own making.

## Preservation and occurrence in sedimentary rocks

Low-angle cross-stratification is in one of the most favorable positions for preservation found along ephemeral streams. In addition to its occurrence in thick units at the bases of sedimentation units, the formation of low-angle cross-stratification on point and longitudinal bars favors preservation. Erosion by subsequent floods is less likely in this position than in any other.

## Other occurrences

Relatively large low-angle cross-stratification occurs widely in deposits of beach, shallow-marine, lacustrine and all fluvial environments. However, in the beds of most of these settings inclinations are greater, on the order of 10–15 degrees rather than 5–10 degrees, and other types of cross-stratification are more abundant. For example, in braided stream deposits high-angle planar and trough cross-stratification are the most abundant types (Williams and Rust, 1969, pp.653–662; Smith, 1970, pp.2998–3000). In meandering river sequences, trough cross-stratification is dominant (Frazier and Osanik, 1961, pp.124–135; Harms et al.,1963,

Fig.122. Avalanche-front cross-stratification in channel bank in tributary of Coyote Wash. Flow is from right to left. This is an example of a bar front slightly modified by current action. The inclination is less than the angle of repose and the strata are concave up. Length of scale is 15 cm.

Fig.123A.

pp.570–576; McGowen and Garner, 1970, p.107). In shallow-marine and lacustrine rocks small-scale cross-stratification is dominant and larger sets tend to be either high-angle or trough (Imbrie and Buchannan, 1965, pp.151–160; Klein, 1965, pp.179–183). Thus, the abundant medium to large-scale, low-angle cross-stratification described here may be diagnostic of shallow or ephemeral streams.

## Significance

Low-angle cross-stratification may be indicative of shallow or ephemeral stream deposits. In fluvial rocks it is restricted to point and longitudinal bars. The maximum inclination is downcurrent or toward the channel. Low-angle cross-stratification probably is deposited by traction from relatively weak currents that approach upper-flow regime conditions for the size of sediment being deposited.

## Selected reference

Williams, G. E., 1966. Planar cross-stratification formed by the lateral migration of shallow streams. *J. Sediment. Petrol.*, 36: 742–746.

Fig.123.A,B. Avalanche-front cross-stratification on composite bar, Coyote Wash. Current is from left to right. Several individual sets are visible. Note sinusoidal ripple stratification with slight upstream drift at foot of fine-grained set in center. The top of the bar is flat and thickness of the upper set varies with the underlying topography. Inclined and horizontal discontinuous stratification at left. Micro cross-stratification and horizontal parallel stratification at surface. Fifteen-centimeter ruler for scale.

## AVALANCHE-FRONT CROSS-STRATIFICATION
(mega ripple bedding, high-angle planar cross-stratification, tabular cross-stratification, torrential cross-stratification)

### Description

Avalanche-front cross-stratification is composed of single tabular sets of thick strata inclined downcurrent at high angles (Fig.122, 123,a,b). The sediment is coarse sand and gravel and the strata are at the angle of repose, between 30 and 40 degrees. The lower surface is planar to irregular, and erosional. Strata are generally straight but some are curved near the base. Sets are approximately 30 cm thick and about 1 m long. Coarsest grains tend to be concentrated near the base.

### Occurrence

Avalanche-front cross-stratification is a minor structure that is developed in transverse and longitudinal bar deposits.

### Origin

Avalanche-front cross-stratification results from sediment slumping into quiet water on the lee face of a bar. In effect it is the foreset beds of a small Gilbert-type delta. Sediment is transported in traction across the top of the bar. At the lee edge, the water deepens appreciably and current velocity decreases, resulting in bedload deposition at the slope break. This deposit increases until it slumps down the bar front under its own weight. Consequently, each stratum is at the angle of repose, is straight, and sharply abuts against the bottom. Where there is some current activity on the lee face, the angle is somewhat less and the strata are slightly concave-upward. The upper surface of the set is horizontal and records the approximate water level. The lower surface follows an erosional contact. Commonly the sets are wedge-shaped as the bar migrates into deeper water. Transportation of the sediment across the bar is by upper-flow regime currents when the water is at least 5 cm deep. Shallower water depths greatly retard current velocity and result in asymmetric ripples. As the flood waters wane, the avalanche fronts probably are deactivated suddenly. The coarse size of the sediment requires vigorous currents for transportation. Moderate currents are capable of only minor winnowing. Thus avalanche-front cross-stratification commonly is followed by non-deposition, and does not show an upward suite of structures indicative of decreasing currents. Where this bedding develops in finer sediment, as it does along braided streams, an upward suite is present (Smith, 1970, pp.2998–3000).

The relative scarcity of avalanche-front bedding possibly is indicative of small or ephemeral streams. Avalanche fronts can develop only where bars are migrating in water several feet deep. In shallower depths, current separation around the bars is incomplete, resulting in degradation of slip faces to angles less than the angle of repose. In the terminology of Bagnold (1941, pp.127, 238), accretion deposits develop rather than avalanche deposits. Only where the water depth is great enough to provide an effective lee-side current shadow do abundant avalanche deposits form.

### Preservation and occurrence in sedimentary rocks

Because of the scarcity of avalanche-front cross-stratification along ephemeral streams we can only speculate as to its possible preservation. The structure is present in sedimentary rocks and it should be preserved in ephemeral stream deposits.

## Other occurrences

Avalanche-front cross-stratification is abundant in the deposits of braided streams, especially in downstream reaches (Smith, 1970, pp.2998–3000). The structure is also common in beds deposited in tidal and shallow marine environments.

## Significance

This type of bedding is not diagnostic of fluvial environments. However, it is indicative of high-velocity currents and relatively deep water. The maximum dip of laminae is downcurrent or oblique to the channel.

## Selected references

Bagnold, R. A., 1941. *The Physics of Blown Sands and Desert Dunes*. Methuen, London, 165 pp.

Harms, J. D. and Fahnestock, R. K., 1965. Stratification, bed forms, and flow phenomena (with an example from the Rio Grande). *Soc. Econ. Paleontologists Mineralogists, Spec. Publ.*, 12: 84–115.

Smith, N. D., 1970. The braided stream depositional environment: comparison of the Platte River with some Silurian clastic rocks, north-central Appalachians. *Bull. Geol. Soc. Am.*, 81: 2993–3014.

Williams, P. F. and Rust, B. R., 1969. The sedimentology of a braided river. *J. Sediment. Petrol.*, 39: 649–679.

# BACKSET CROSS-STRATIFICATION

## Description

Backset cross-stratification is characterized by upcurrent dips. In ephemeral streams care must be exercised to differentiate between true backset cross-stratification and cross-stratification that dips upstream (not upcurrent) because of channel sinuosity, local flow reversal and so forth. In trenches it is not uncommon to observe numerous sets inclined roughly upstream. For example, in one large trench at Twelvemile Wash, the average direction of fourteen cross-stratified sets was 145° from the channel direction. However, most of the readings were from point-bar deposits and the anomalous directions result from channel migration, eddy currents and inclined stratification. To confirm the occurrence of backset cross-stratification, an independent indicator of current direction is needed within the same unit. Pebble imbrication was used in this bed (Fig.124).

Backset cross-stratification is developed in coarse, pebbly sand. Strata are coarse and form tabular sets several centimeters thick. Angles of inclination are

Fig.124. Backset cross-stratification in channel deposit of tributary to Coyote Wash. Current is from right to left. Note the pebble imbrication. Several tabular planar sets with low-angle, upstream dipping strata are present. Fifteen-centimeter ruler for scale.

low, generally less than 10° upcurrent. Strata are straight and discordant and the lower set boundary is planar. Several sets are superimposed.

## Occurrence

Backset cross-stratification is rare in ephemeral stream deposits. The occurrence at Coyote Wash (Fig.124) was the only one confirmed, but several other suspected examples were observed. Too few were seen to suggest generalizations as to mode of occurrence.

## Origin

Backset cross-stratification results from high-velocity currents in the upper-flow regime (Jopling and Richardson, 1966). The low-angle type observed along ephemeral streams is intermediate between lower-velocity planar beds and higher-velocity high-angle backsets (Jopling and Richardson, 1966, p.821). These bedding types correspond to plane bed, antidune, and chute and pool bedforms, respectively (Simons and Richardson, 1963, pp.321, 322). Considering the coarse grain size, associated horizontal discontinuous stratification, lack of ripple marks and analogies with numerous flume experiments, the backset cross-stratification observed at Coyote Wash is interpreted to have formed from antidunes under upper-flow regime conditions.

### Preservation and occurrence in sedimentary rocks

Backset cross-stratification is likely to be preserved in deposits of ephemeral streams, but not abundantly. The scarcity of this structure in sedimentary rocks probably reflects actual scarcity combined with lack of criteria for recognition of the structure.

### Other occurrences

Hand et al. (1969) and Picard (1970b) have noted the natural occurrences of antidunes in alluvial fans, ephemeral streams, beaches, braided rivers and floodplains. However, too little is known of structures produced by antidunes to estimate their relative abundance in beds of various environments.

Fig.125. Transverse section of flow rolls in point bar deposit, Little Mountain Creek. Low-angle cross-stratification is present above and below the disturbed zone. Pen is 13.5 cm long.

Fig.126. Flame structures at top of ripple-stratified set in point-bar sedimentation unit, Little Mountain Creek. Pen is 13.5 cm long.

Fig.127. Thick zone of convolute stratification at top of longitudinal bar deposit, Halfway Hollow. Disturbed unit is overlain by floodplain deposits.

## Significance

Backset cross-stratification is not diagnostic of environment of deposition. It forms under upper-flow regime conditions and the strata are inclined upcurrent.

## Selected references

Hand, B. M., 1969. Antidunes as trochoidol waves. *J. Sediment. Petrol.*, 39: 1302–1309.
Hand, B. M., Wessel, J. M. and Hayes, M. O., 1969. Antidunes in the Mount Toby Conglomerate (Triassic), Massachusetts. *J. Sediment. Petrol.*, 39: 1310–1316.
Jopling, A. V. and Richardson, E. V., 1966. Backset bedding developed in shooting flow in laboratory experiments. *J. Sediment. Petrol.*, 36: 821–825.
Middleton, G. V., 1965. Antidune cross-bedding in a large flume. *J. Sediment. Petrol.*, 35: 922–927.
Power, W. R., Jr., 1961. Backset beds in the Coso Formation, Ingo County, California. *J. Sediment. Petrol.*, 31: 603–607.
Simons, D. B. and Richardson, E. V., 1963. Forms of bed roughness in alluvial channels. *Trans. Am. Soc. Civil Eng.*, 128: 284–323.

## DISTURBED STRATIFICATION, I. SLUMP FEATURES: CONVOLUTE STRATIFICATION, FLAME STRUCTURE, FLOW ROLL

(slump roll, ball-and-pillow, pillow structure, core-and-shell, pseudo-nodule, storm roller, flow structure)

## Description

Slump features are post-depositional disturbances of original bedding. Convolute stratification, flame structures and flow rolls were observed along the ephemeral streams studied (Fig.125–127). All of these bedding types preserve the original strata virtually intact, but the attitude of the stratification is altered. Convolute stratification is a general name that encompasses many slump features. This structure is characterized by unbroken strata that are bent into a series of folds, ranging from open undulations to recumbent isoclinal folds. Generally, all stratification within the set is disturbed and the set is bounded by sharp planar surfaces. Sediments above and below the convolute set rarely are disturbed. The size of folds is variable; the amplitude of most folds observed was 5–15 cm. The thickness of the convolute set generally is only slightly greater than fold amplitude; multi-story folding within a single set is rare.

In flow rolls the strata are broken and short segments are rolled up like a carpet (Fig.125). Strata are concentric around axes of rolls. In cross-section, flow rolls are circular to oval in shape; in longitudinal section they are rod or spindle shaped. Long axes of flow rolls tend to be aligned subparallel with each other.

Flame structures are upright, peaked, asymmetric folds that tend to develop in regularly spaced trains (Fig.126). Most of them are overturned at the top in one direction, giving the appearance of breaking waves. The general shape is that of a steep, isosceles triangle, the top of which is skewed or bent over. With extreme bending, the structure approaches a recumbent isoclinal form.

**Occurrence**

Slump features occur extensively at the top of buried bar sequences. In virtually all point and longitudinal bar sedimentation units that were covered by younger sediment, the rippled zone near the top contained slump features. In contrast, where ripple stratification is replaced by micro cross-lamination, slump features are rare. Within the same sedimentation unit a disturbed set commonly contains many types of slump features and grades laterally into ripple stratification. Slump features also occur in low-angle cross-stratification but, as with ripple stratification, only at the top of the unit.

**Origin**

Slump features result from downslope movements of the overlying sediment. Movement occurs at the interface between sliding cover and immobile basement and friction produces a stress couple. Thin sets of ripple stratification that are developed at the top of buried units apparently are sufficiently weak so that stress is released and the strain is confined largely to these horizons. Where ripple stratification is absent, the strata at the interface generally are strong enough to resist the weak stresses induced by the overlying sediment load and the slight depositional slope.

An origin by slumping, rather than compaction and foundering, is indicated by the downslope orientations of flame structures and flow rolls (the latter occur as both rollers and streamers, but rarely oblique) and the lack of compaction features such as faults, thixotropic effects, and pipes for escaping fluids. Only slight slumping is indicated by the essentially undisrupted fabric, although truncated structures at the upper contact show some detachment and actual slippage of cover over basement. Proof that slump features are post rather than syn-depositional is seen by the lack of such structures in sedimentation units that are not buried and their occurrence in virtually all buried units.

**Preservation**

Although slump features are developed in thin zones at the tops of sedimentation units (a setting generally not conducive to preservation), the chances of preservation for flow structures are good. Inasmuch as the structure forms only after burial and is present in point and longitudinal bar deposits, erosion by subsequent floods is less likely than with many other types of bedding.

### Other occurrences

We and many others have observed similar features in fluvial, deltaic, shallow-marine, deep-marine and lacustrine deposits (Cooper, 1943; Fairbridge, 1947; Emery, 1950; Sanders, 1960; Stewart, 1963; Selley et al., 1963; Grant-Mackie and Lowrey, 1964; Davies, 1965; Kelling and Williams, 1966; Spreng, 1967; Lowry and Cooper, 1970; Morris, 1971).

### Significance

Slump features are not diagnostic of environment of deposition. In fluvial rocks they are restricted to bar deposits and indicate the downslope direction.

### Selected references

McKee, E. D., Reynolds, M. A. and Baker, G. H., 1962a. Laboratory studies on deformation in unconsolidated sediment. *U.S. Geol. Surv., Profess. Papers*, 450D: 151–155.

McKee, E. D., Reynolds, M. A. and Baker, G. H., 1962b. Experiments on intraformational recumbant folds in cross-bedded sands. *U.S. Geol. Surv., Profess. Papers*, 450D: 155–160.

Sorauf, J. E., 1965. Flow rolls of Upper Devonian rocks of south central New York State. *J. Sediment. Petrol.*, 35: 553–563.

## DISTURBED STRATIFICATION, II. BIOTURBATION: BURROWS
(churned stratification, sublenticular or wispy stratification)

## ROOT DISRUPTED
(mottled stratification, gleyed, paleosol, sublenticular or wispy stratification)

### Description

Bioturbation structures are post- and syn-depositional deformed strata resulting from organic activity. Included are burrowing and root growth. Tracks and footprints could also be included because large animals can destroy original bedding to depths of several inches, especially where large numbers congregate (feeding grounds, water holes).

Burrows range from small, delicate feeding and habitation structures excavated by invertebrates to relatively large tunnels dug by reptiles, rodents and

Fig.128. Animal burrows and ripple stratification in overbank deposits, Kennedy Wash.

larger mamals. The tunnels made by invertebrates can be so extensively developed as to completely disrupt the original fabric and even destroy the distinctive burrow morphology.

**Occurrence**

Along ephemeral streams, bioturbation structures generally are limited to floodplains. The only exceptions are isolated invertebrate burrows in channel deposits and pits dug by animals searching for water. Animal pits were found only at Coyote Wash where they are rare. Burrows on the floodplain are widely spaced and do not lead to much disturbance of the strata (Fig.128). In contrast, disruption by plant roots can be extensive.

**Origin**

The origin of bioturbation structures is evident. The principal problem concerns the interpretation of extensively disturbed strata and the identification of the agents. Along ephemeral streams, sublenticular or wispy stratification was observed to result only from plant activity. The density of animal burrows is not great enough to produce such extensive disturbance nor are other physical modes of disturbance that were observed. The identification of buried soil profiles is equally simple because of the presence of abundant macerated plant remains and the underlying root zone. However, in ancient rocks or in other environments both of these problems are much more difficult than they are in ephemeral stream settings.

### Preservation and occurrence in sedimentary rocks

Because so little is known about the processes that operate on floodplains, it is difficult to evaluate the potential for preservation of bioturbation structures. In an aggrading stream system, the preservation of floodplain features is likely. Preservation of structures outside the floodplain on highland surfaces is less likely.

### Other occurrences

Ancient soil horizons are restricted to fluvial and deltaic environments. In contrast, burrowing occurs in almost all environments, but is most pronounced in shallow-marine settings.

### Significance

Soils are diagnostic of fluvial floodplains, upland surfaces, and stable conditions.

### Selected references

Grim, R. E. and Allen, V. T., 1938. Petrology of the Pennsylvanian underclays of Illinois. *Bull. Geol. Soc. Am.*, 49: 1485–1513.

Reiche, P., 1950. A survey of weathering processes and products. *Univ. New Mexico Publ. Geol.*, 3: 95 pp.

Termier, H. and Termier, G., 1963. *Erosion and Sedimentation*. Van Nostrand, London, 433 pp.

## DISTURBED STRATIFICATION, III. DESICCATION FEATURES: MUD PEBBLE (intraformational or flat clast)

## DESICCATION CRACK

### Description

Desiccation features are disturbed strata that result from the evaporation of interstitial water. Cracks are the most common form and were discussed earlier.

Fig.129. Remnants of once continuous mud film in channel deposit, Coyote Wash. Only isolated pebbles remain after partial erosion of mud interval. Scale is in inches and centimeters.

In cross-section, desiccation cracks are V-shaped openings that penetrate the sediment to depths of several centimeters. When buried, desiccation cracks are recognizable because of the disruption of strata and infilling by overlying sediment. Associated with cracks is the upward curling of laminae, largely affecting horizontal parallel stratification (Fig.110). Mud pebbles are formed when fragments of cracked layers are redeposited in younger sediment (Fig.116). Partial erosion of mud films can also leave isolated mud pebbles (Fig.129).

**Occurrence**

Desiccation cracks rarely are encountered below the surface. Mud pebbles are only slightly more abundant and are found most often in channel deposits.

**Origin**

The origin of desiccation features was discussed under the heading of various types of cracks. To be preserved as stratification features, cracks must be infilled by sediment from the bed above without significant scour. The preservation of clay pebbles requires short transportation and redeposition in younger sediment.

### Preservation and occurrence in sedimentary rocks

The wide disparity in abundance between desiccation cracks on the surface and in the subsurface indicates the extreme difficulty of preservation of this structure. Although produced in abundance, desiccation cracks are too delicate and shallow to survive, except rarely. Similarly, mud pebbles are not strong enough to commonly survive transportation and redeposition by the vigorous currents of ephemeral streams.

### Other occurrences

Desiccation features are present in fluvial, tidal, marginal marine and lacustrine beds.

### Significance

Desiccation features are not diagnostic of environment of deposition. Mud pebbles are most abundant in channels. Desiccation features are indicative of fluctuating water levels.

## DISTURBED STRATIFICATION, IV. COMPACTION FEATURES: COMPACTION FAULTS
(penecontemporaneous faults)

### Description

Compaction faults are high-angle normal faults that form singly or in sets. Displacement generally is less than 3 cm and stratification is otherwise undisturbed. Fault penetration can be a meter or more.

Thixotropic structures, although not observed in this study, should occur in association with quicksand, which was encountered. Expected thixotropic features include completely disturbed stratification, clastic dikes and injection pipes.

### Occurrence

Compaction faults occur rarely in fine sands of point-bar deposits. Faults in other sedimentation units were not observed. Quicksand and, presumably, thixotropic features occur on bars and in channels along many streams. The distribution of quicksand is more widespread than is indicated by associated structures (sand volcano, gas bubble). Because of the nature of the sediment, no trenches were dug to determine stratification types associated with quicksand.

## Origin

Differential compaction in fine sand results in small faults. The weight of sand compacts the underlying sediment, in which slight vertical shifts cause passive foundering of the overlying sand. Differential movements are slight and the disturbance is confined to fracture planes. As indicated by the presence of faults in surficial deposits, most of the faulting probably takes place prior to burial.

Thixotropic structures result from increase in trapped pore fluids as overburden is added. Water-saturated sediments under pressure are unstable and are subject to lateral or upward migration to release abnormal interstitial pressures. As a result of the movement, strata are disrupted within and above the set.

## Preservation and occurrence in sedimentary rocks

Although only a few examples were found, compaction-disturbed stratification should be preserved as a minor structure in ephemeral stream deposits.

Fig.130. Stepped erosion surfaces and mud films separating point-bar units, Twelvemile Wash. Mud films separate six point-bar sedimentation units. The uppermost unit and the three lowermost units are complete. The second and third units from the top (below stepped surfaces) have had most of the sequence removed. Note the convolute bedding in the second unit from the bottom, just above the thick ripple-stratified layer. Low-angle cross-stratification is predominant. Note the reversal of inclination in the upper unit. Flow is from right to left. View toward channel.

## Other occurrences

Compaction faulting occurs widely in many environments. However, the structure is probably most abundant in fluvial deposits because of the extreme lateral changes in sediments resulting in differential compaction. In marine environments, conditions and sediments are more uniform over local areas and compaction would tend to be more nearly equal.

Thixotropic structures are present in many fine-grained deposits. In sands, the structures probably are limited to areas of rapid deposition, such as deltas, streams and beaches.

## Significance

Compaction features are not diagnostic of environments of deposition.

## Selected reference

Boswell, P. G. H., 1961. *Muddy Sediments*. W. Heffer, Cambridge, 140 pp.

# DISTURBED STRATIFICATION, V. EROSIONAL FEATURES: STEPPED EROSION SURFACE

## Description

Erosional surfaces that transect older stratification can be considered as a form of disturbance. Planar, trough and irregular erosional surfaces between and within sedimentation units have been described here in conjunction with other types of bedding. Some of the surfaces between sedimentation units show a distinctive stepped profile (Fig.130). The steps have long, gently sloping treads and vertical or overhanging risers less than an inch in height. Mud films generally are draped over the surface.

## Occurrence

Stepped erosion surfaces are common between sedimentation units in point-bar deposits; none were observed in channel deposits. Mud films are associated with the majority of the erosion surfaces.

## Origin

Stepped erosion surfaces are buried micro-terraces (Fig.28, 29). Small terraces are cut by waves on point bars emerging from flood waters. Subsequent burial preserves the erosion surface. The associated mudfilm originates, not from the flood that deposits the point bar, but from subsequent weaker floods. These later floods are too small to do more than scour the face of the bar, depositing a mud mantle across the channel. Thus, the point bar shown in Fig.130 is a composite feature of six major floods (point-bar construction) and at least five minor floods (intervening mud films).

### Preservation and occurrence in sedimentary rocks

As shown in Fig.130, preservation of stepped profiles and mud films is likely. In sedimentary rocks, the topographic relief on the erosion surface might be so modified by compaction that it appears to be only irregular, rather than stepped; mud films would be preserved as thin shale interbeds.

### Other occurrences

Stepped erosion surfaces are widespread in bar deposits along all types of streams, although perhaps not in the abundance and detail of those noted in ephemeral streams. The structure can also be expected in tidal flat deposits.

### Significance

If they are abundant, stepped erosion surfaces possibly are suggestive of ephemeral stream deposition. They are indicative of fluctuating water levels and changing conditions.

### Selected reference

Ore, H. T., 1964. Some criteria for recognition of braided stream deposits. *Contrib. Geol.*, 3: 1–14.

*Chapter 6*

# SEDIMENTATION UNITS

Although elusive in theory, the concept of sedimentation units has proven to be useful in practice. As used here, a sedimentation unit is defined as one or more sets of strata deposited continuously under uniform or continuously varying conditions. A sedimentation unit thus represents a single episode of sedimentation. This usage is consistent with the original definition of Otto (1938, p.575) and agrees with Pettijohn's interpretation (1957, p.159).

Although defined thirty-four years ago, sedimentation units have only recently received much attention and the literature on this subject is still sparse. Many of the theoretical difficulties of this concept are discussed by Jopling (1964) who points out the problems with scale, duration, and continuity. While we do not wish to minimize the questions raised by Jopling, these problems are not significant in many, but not all, applications, and sedimentation units are being used increasingly to reconstruct ancient depositional processes.

Sedimentation units are a special type of cyclic sequence. Sedimentary cycles can be classified as either allo- or autocyclic. In the former the cyclic pattern is in response to external processes, such as uplift or eustatic change, while in the latter the cycles develop from processes intrinsic to the depositional environment, such as channel migration or delta progradation. As defined here sedimentation units are a special case of autocyclic sequences. Precise definitions that distinguish autocyclic sequences and sedimentation units are not possible because of the semantic problems discussed by Jopling. However, several examples will make clear the differences.

The vertical sequence resulting from river meanders is well-known as a characteristic fining-upward cycle. Lag channel gravel is overlain by thick, cross-stratified point-bar sand that is in turn overlain by floodplain mud. This pattern has been so thoroughly documented in both recent and ancient deposits as to become the classic example of autocyclic sequences. However, the meandering sequence is not a sedimentation unit. Although the deposits may have accumulated under conditions which, over the long term trended steadily from channel to floodplain, deposition was certainly not continuous. Rather, deposition in a meandering stream is discontinuous and the entire sequence is built up by numerous events, consisting of both individual floods and longer periods of uniform flow.

Thus, sedimentation units are present within this autocyclic sequence and will generally be recognized as sets of conformable strata. Within the channel and bar facies, erosion surfaces are the key to differentiating individual sedimentation units; in the floodplain facies individual laminae may, in addition to sets deposited by overbank flow, also represent sedimentation units.

The second example of sedimentation units is turbidites. In this case the autocyclic sequence and the sedimentation unit are equal. The familiar flysch sandstone facies is an autocyclic unit deposited as a single episode and is therefore also a sedimentation unit. Fine-grained rocks between individual turbidites represents slower, "normal" deposition in the basin. Although deposition may have been continuous, it was more probably episodic, possibly controlled by smaller storms that did not trigger turbidity flows.

A more extended discussion of sedimentation units is premature at this time. We do not intend to resolve any of the ambiguities concerning the concept, but only to describe and illustrate the units occurring in ephemeral stream deposits.

## SEDIMENTATION UNIT I: POINT-BAR SEQUENCE

### Description

Point-bar sedimentation units have the simplest pattern of stratigraphy and sedimentary structures encountered in ephemeral stream deposits. The sequence consists of a fining-upward cycle of low-angle cross-stratification, micro cross-stratification, ripple stratification, horizontal parallel stratification, and a mud film. The lowermost unit of low-angle cross-stratification generally comprises more than one-half of the thickness; other sets are relatively thin. Common modifications to this pattern are: poor separation of micro cross-stratified and ripple-stratified units, replacement of ripple stratification by convolute bedding, lack of horizontal parallel stratification, and partial erosion of upper units.

### Occurrence

The point-bar sequence is common along ephemeral streams (Fig.131, 132), and comprises the major part of all point-bar deposits. Toward the channel, the sequence can be truncated, or it may grade into inclined, channel-fill deposits.

### Origin

The point-bar sequence is deposited as a unit during the waning stages of floods. With the onset of deposition at some time after the passing of the flood crest,

Fig.131. Point-bar sequence, Twelvemile Wash. This photograph is a close up of the uppermost unit shown in Fig.130. Low-angle cross-stratification at the base, ripple stratification above. Scale is 15 cm long.

low-angle cross-strata are deposited while flow in the channel is still vigorous. Upper-flow regime conditions in the channel are probable because of the gradation of low-angle cross-stratification into horizontal discontinuous stratification. Although lower flow velocities occur over the bar, the lack of ripple marks indicates the dominance of upper-flow regime conditions for the finer bar sands also. These conditions persist until the flood has diminished. As velocity decreases and water level drops, the bar is subjected to a brief period of lower-flow regime conditions and cuspate ripple marks may produce micro cross-stratification. This phase is transitory and ends as the water over the bar becomes too shallow for currents.

Fig.132

Fig.133

Linear asymmetric ripples form on the barely awash bar yielding ripple stratification. If the water level drops rapidly enough, the cuspate-ripple stage can be bypassed. If the bar remains awash while flow velocity in the channel continues to decrease, horizontal parallel stratification, parting lineation and a mud film are deposited. However, the water level generally drops before these sets are deposited and the sequence usually ends with ripple stratification.

### Preservation and occurrence in sedimentary rocks

Point-bar sequences are readily preserved, in whole or in part, along ephemeral streams and are common in fluvial rocks (Harms, 1966; Allen, 1970a; Moody-Stuart, 1966; Siever, 1951). Fig.133 shows such a sequence from the fluvial Duchesne River Formation (Eocene-Oligocene?) in northeastern Utah. For good accounts of textural relations, papers by Folk and Ward (1957) and Visher (1969) are recommended.

### Other occurrences

Similar sequences of sediment are deposited in tidal channels. Bipolar paleocurrent patterns are sufficient to distinguish tidal channels from fluvial channels (Klein, 1965, pp.179–186).

### Significance

Sedimentation unit I is diagnostic of fluvial point bars. The sequence is deposited by waning currents. Directional significance of the structures is questionable.

### Selected references

Allen, J. R. L., 1965c. A review of the origin and characteristics of recent alluvial sediments. *Sedimentology*, 5: 89–191.

---

Fig.132. Point-bar sequence over channel deposits, Little Mountain Creek. The poorly stratified sand at the base of the hammer grades into cobble-sized channel gravel a meter to the left. Upper surface of sand layer has several desiccation cracks. Two point-bar sequences are present above the sand ledge. Erosion has removed all but a remnant of the rippled layer above and to the left of hammer head (note surface to the right and below the pebble). The upper unit terminates with micro cross-stratification.

Fig.133. Point-bar sequence in Duchesne River Formation (Eocene-Oligocene?). Lower three-quarters of unit is low-angle cross-stratification. Ripple stratification is present above, and there is a thin layer of horizontal parallel stratification at the top. The upper surface is marked with parting lineation. Length of pen (upper left) is 13.5 cm.

Fig.134. Longitudinal section of channel-bar sequence, Coyote Wash. Flow is from left to right. The channel-bar sequence has modified avalanche cross-stratification at the base which is overlain by horizontal discontinuous stratification capped by micro cross-stratification. Above the channel-bar unit there are two additional sedimentation units that consist of horizontal discontinuous stratification (channel fill) and low-angle cross-stratification (point bar). Exposed part of scale is about 11 cm.

Allen, J. R. L., 1970a. Studies in fluviatile sedimentation: a comparison of fining-upwards cyclothems, with special reference to coarse-member composition and interpretation. *J. Sediment. Petrol.*, 40: 298–323.

Bernard, H. A., Major, C. F. Jr., Parrott, B. S. and LeBlanc, R. J., Sr., 1970. Recent sediments of southeast Texas. *Bur. Econ. Geol., Univ. Texas, Austin, Guidebook*, 11.

Klein, G. de V., 1965. Dynamic significance of primary structures in the Middle Jurassic Great Oolite Series, southern England. *Soc. Econ. Paleontologists Mineralogists, Spec. Publ.*, 12: 173–191.

McGowen, J. H. and Garner, L. E., 1970. Physiographic features and stratification types of coarse-grained point bars. *Sedimentology*, 14: 77–111.

## SEDIMENTATION UNIT II: CHANNEL-BAR SEQUENCE

### Description

Channel-bar sequences are the deposits of longitudinal and transverse bars within

Fig.135. Channel-bar sedimentation unit, Cliff Creek. Transverse section. Flow is away from viewer. The channel-bar sequence occupies the lower half of the face. It consists of modified avalanche-front cross-stratification at the base, inclined horizontal discontinuous stratification in the center, and micro cross-stratification at the top (in dark band near the top of scale). Above the top of the scale (14 cm exposed) there is a point-bar sequence, low-angle cross-stratification, micro cross-stratification, and horizontal parallel stratification.

channels (Fig.134–136). These sequences are slightly more complicated and variable than point bar units. The general sequence consists of avalanche-front cross-stratification at the base topped by a thin layer of horizontal discontinuous stratification. The sequence can be terminated at this point or it may continue with thin units of festoon cross-stratification, micro cross-stratification and horizontal parallel stratification. Channel-bar sequences are thinner than point-bar units and rarely exceed 0.3 m in thickness. Point-bar deposits are commonly superimposed on channel-bar sequences (Fig.134–136).

## Occurrence

Channel-bar sequences are not as abundant as either point-bar or channel-fill deposits. This sedimentation unit is common in ephemeral stream deposits, but is subordinate to other types.

Fig.136. Longitudinal section (same position as Fig.135). Flow is from right to left. The top of the channel-bar sequence is about 5 cm above the top of scale. Horizontal discontinuous strata overlie slightly lenticular horizontal discontinuous strata. The micro cross-stratification at the top of the channel-bar sequence has been eroded. Length of scale is 15 cm.

## Origin

The lower two units, avalanche-front cross-stratification and horizontal discontinuous stratification, record an advancing channel bar. The layers represent, respectively, the foresets and topsets of a Gilbert-type delta. On the top of the bar, shallow-water and high-flood discharge result in high current velocities and upper-flow regime conditions. Sediment is transported continuously along the bed and forms streaming lineation, lag ripples and horizontal discontinuous stratification. Sediment carried to the front of the bar is deposited as avalanche strata at the angle of repose if the water is deep enough, or as modified avalanche strata at lesser angles if the water is shallow. Thus, the sequence at any one point is from foresets, filling relatively deep and quiet water, to horizontal discontinuous stratification in shallow water. If the bar remains under water while the flood decreases, a series of units indicative of diminishing current velocity are deposited. Deposition ceases when the bar is exposed by falling flood waters.

## Preservation and occurrence in sedimentary rocks

Channel-bar sequences are found in sedimentary rocks and are readily preserved in deposits of ephemeral streams.

## Other occurrences

Channel-bar sequences are present in deposits of braided streams (Mrakovich, 1969), but cycles showing an upward decrease in current velocity are much more common (Ore, 1964, p.10; Smith, 1970, pp.2994–2995, 2999). Similar cycles also are developed in tidal channels, but point-bar sequences are more abundant (Klein, 1965, pp.184–186).

## Significance

Channel-bar sequences are suggestive of fluvial environments. Possibly the channel-bar sequence described here is indicative of ephemeral conditions.

In coarse bed streams braiding is caused by the construction of low, linear, mid-channel mounds during periods of high discharge (Smith, 1971b). The formation of these longitudinal mounds requires only that a stream at some point become unable to transport part of its coarsest load (Leopold and Wolman, 1957). Longitudinal bars are dominant in the upper reaches of the South Platte River in Colorado (Smith, 1970) and in many other coarse bed streams (Doeglas, 1962; Fahnestock, 1963; Ore, 1964; Williams and Rust, 1969). In contrast, braided streams that are sand-dominated are characterized by transverse bars (Collinson, 1970; McGowen and Garner, 1970; Culbertson and Scott, 1970; Waechter, 1970; Smith, 1971b). Apparently, transverse bars form by sediment aggrading to a profile of equilibrium (Jopling, 1966; Smith, 1971) and grow by downcurrent extensions of avalanche faces. Depth, velocity and grain size tend to decrease on active bar surfaces from their upstream mouths to the downstream and lateral margins (Smith, 1971b).

## Selected references

Doeglas, D. J., 1962. The structure of sedimentary deposits of braided streams. *Sedimentology*, 1: 167–170.

Smith, N. D., 1970. The braided stream depositional environment: comparison of the Platte River with Silurian clastic rocks, north-central Appalachians. *Bull. Geol. Soc. Am.*, 81: 2993–3014.

Smith, N. D., 1971b. Transverse bars and braiding in the Lower Platte River, Nebraska. *Bull. Geol. Soc. Am.*, 82: 3407–3420.

Fig.137. Channel-fill sedimentation unit, Cliff Creek. Flow is from left to right. Horizontal discontinuous strata at base (lower surface of unit not seen). The slight inclination on the right represents the influence of a nearby point bar. This layer is continuous with low angle cross-stratification on the bar. Thin units of micro cross-stratification and horizontal parallel stratification are present at the top. Note several burrows at top center. Length of scale is 15 cm.

SEDIMENTATION UNIT III: CHANNEL-FILL SEQUENCE

## Description

Channel-fill sequences closely resemble point-bar deposits. From the base upward, the cycle consists of: basal erosion, lag deposit, inclined or horizontal discontinuous stratification, festoon cross-stratification, micro cross-stratification, and horizontal parallel stratification. The progression commonly is terminated at some point midway through the cycle by non-deposition or erosion. Nowhere was a complete cycle observed. In addition to truncation, actual channel-fill deposits tend to be characterized by preferred development of either the basal members or the upper members. Thus, cycles that were observed can be characterized as dominated by coarse lag deposits (Fig.107) or sands (Fig.108). On a large scale, both patterns are lenticular and channel deposits are interbedded with bar and floodplain units (Fig.137). Identification of coarse lag deposits as channel-fill units is relatively

Fig.138. Gravel-dominated channel-fill units with bar and floodplain deposits, Halfway Hollow. Note the slight imbrication in the gravel and the convolute bedding to the left of the hammer head.

simple; most of them show imbrication and lenticular geometry. The problem is more difficult with sand, but channel fills generally are characterized by inclined stratification and festoon or micro cross-stratification in place of the low-angle cross-stratification and ripple stratification of bars. Also, channel-fill deposits lack stepped erosion surfaces, and mud films are rare.

**Occurrence**

Channel-fill units are widespread and abundant. Gravel-dominated patterns are most common at the proximal end and sandy units are dominant downstream

Fig.139. Sand-dominated channel-fill sequence, Twelvemile Wash. Flow is from right to left. Three sedimentation units are present. The lowest (dark sediment) consists of ripple stratification. Patches of macerated plant material give this unit a sublenticular appearance. High organic content in the upper part imparts the black stain. This surface was exposed, partially oxidized, and developed desiccation cracks. The middle sedimentation unit consists of horizontal discontinuous stratification below and ripple and convolute stratification above. The upper unit is on a marked erosional base; the channels are filled with inclined strata. Length of pen is 13.5 cm.

(Fig.138). Where stream banks expose large areas of alluvium, channel-fill deposits apparently are approximately equal in volume with point-bar and floodplain deposits; channel-bar sequences are subordinate.

## Origin

The channel-fill sequence records erosion by the waxing flood and backfilling as the flood decreases (Fig.139). The influences of decreasing current velocity are best seen by the vertical trend in sedimentary structures displayed by the sandy pattern. Inclined or horizontal discontinuous strata are first deposited while the flood waters are high and upper-flow regime conditions prevail. As the flow decreases with the flood, dunes (festoon cross-stratification), cuspate ripples (micro cross-stratification), and parting lineation (horizontal parallel stratification) are successively produced.

## Preservation and occurrence in sedimentary rocks

Preservation of channel-fill deposits is likely and these units should be present in abundance in sedimentary rocks.

## Other occurrences

Similar sequences are present in the Old Red Sandstone of Scotland (Bluck, 1967, pp.154–159). The bedding is broadly lenticular and channel deposits alternate with floodplain deposits. Conglomerate is dominant at proximal localities and sandstone at distal localities. Similar relations were observed in the Duchesne River Formation. Many of the references cited previously in this chapter are also of interest with regard to channel-fill sequences.

## Significance

Channel-fill sequences are diagnostic of fluvial deposits. Quite possibly channel-fill units of the type described here are restricted to ephemeral conditions.

## Selected references

Allen, J. R. L., 1970c. A quantitative model of grain size and sedimentary structures in lateral deposits. *Geol. J.*, 7: 129–146.

Bluck, B. J., 1967. Deposition of some Upper Old Red Sandstone conglomerates in the Clyde area. A study in the significance of bedding. *Scot. J. Geol.*, 3: 139–167.

*Chapter 7*

# SUMMARY AND CONCLUSIONS

## SEDIMENTARY FEATURES

Observations at several dozen sites along ephemeral streams in the Uinta Basin, Utah, have led to the recognition of 42 different sedimentary structures (or varieties) and 14 major bedding types. Although the mere recognition of these structures is of limited significance in itself, the occurrence of such a variety suggests that the structures are sensitive to local conditions and are useful for reconstructing depositional environments. We realize that this book does not represent an exhaustive description of all of the sedimentary features that are present along ephemeral streams. Undoubtedly other structures are present and are familiar to some readers. Nevertheless we believe that the structures described here are representative and form a firm basis for further discussion.

Although some of the structures that we have described are rare, few of them are insignificant. All of them have some potential for yielding useful paleo-environmental interpretative information. Indeed, many of the minor structures may be more valuable in reconstructing ancient environments than the more abundant features, inasmuch as their scarcity may reflect restricted or even unique conditions at the time of deposition. On the other hand, many minor structures may be so labeled only because of our lack of knowledge concerning their abundance and distribution. These structures may be more obscure or less striking than associated features and therefore pass unrecognized. Two examples are patterned cones and linear-shrinkage cracks. The former structure was described only recently by Boyd and Ore (1963). Since the publication of their original description we have found this structure in a wide variety of sedimentary formations. Linear-shrinkage cracks have long been known from ancient rocks and have been variously attributed to trails of organisms and desiccation. We first observed this structure in the Green River Formation, but have subsequently noticed it in a number of other units. Despite its scarcity in the geologic literature, linear-shrinkage cracks are more common than is generally assumed.

A separate problem concerns the abundant sedimentary structures. It is frequently assumed that because these structures are abundant they are relatively insensitive to environment of deposition and are therefore of limited use in interpreting ancient environments. Examples include shrinkage cracks, ripple marks, and cross-stratification, each of which is considered to form in many sedimentary environments under a broad range of physical conditions. Although true, this viewpoint ignores the potential of these structures for detailed environmental reconstructions. This potential is fully recognized only when these large classes of sedimentary structures are divided into their many varieties. For example, we may cite the many varieties of ripple marks observed along ephemeral streams in the Uinta Basin. We have described seven varieties of solitary ripple marks and five varieties of multiple sets or interference ripple marks. Although the recognition of this many different types of ripple marks may strike some readers as excessive splitting, it is necessary to work at this level of description if the structure is to be useful for detailed reconstructions of depositional environments. It is one of the major goals of this book to demonstrate that these many varieties do exist as discrete structures that are easily recognizable in the field. We have not resorted to detailed morphometric analysis in order to differentiate between the several varieties. Although gradations exist between one type and another, it is generally possible to classify any one structure on the basis of field observation.

## ORIGIN AND OCCURRENCE

Many factors combine to control the distribution of sedimentary structures along ephemeral streams. These factors include: sedimentary properties such as grain size, sorting and composition; hydrodynamic properties such as sediment load, discharge, width, depth, flow velocity, bed roughness and flood duration; stream pattern and slope; position within the fluvial environment; organic activity; and time. None of these variables (except possibly load and discharge: see, Leopold and Maddock, 1953) are strictly independent, and it is impossible to reconstruct the actual interrelations that occur in natural systems. Nevertheless it is possible to isolate several factors that are significant in directly producing many common sedimentary structures. These significant factors include grain size, subenvironment, and flow velocity. Of these three factors, water velocity is by far the most significant and directly controls the occurrence of many types of sedimentary structures. Grain size exerts a positive but crude control. The effects of subenvironment are relatively obvious.

Downstream changes in width and depth of a stream channel are greatly influenced by sediment type, but little quantitative information is available (Schumm, 1960a,b, 1961, 1963, 1968). Apparently the role of grain size in determining the occurrence of sedimentary features is also complex. Although it can be argued that in any fluvial system grain size and climate are the independent variables that determine other parameters such as slope, flow velocity, width, depth, and so forth, the control exerted by grain size is not direct but operates through the semi-dependent or dependent variables. The direct control exerted by grain size appears to be fairly elementary. The occurrence of many structures requires sediment of specific caliber. For example, imbrication requires a high concentration of gravel in the sediment, and crescent scours require a sparse concentration. Ripple marks and cross-stratification are best developed in sand or coarse silt; rarely are they formed in gravel or mud. Similarly, quicksand features are also mostly restricted to sand and coarse silt. In contrast, thixotropic structures originate in fine-grained sediment. Beyond this level of interpretation, it is questionable whether grain size exerts any direct control on the occurrence of sedimentary structures. Table XIII lists the results of size analyses for several different types of sedimentary structures. Although there is a correlation between grain size (as indicated by percent of mud) and the type of sedimentary structure, this pattern may be more apparent than real, reflecting not a cause and effect relationship but a mutual dependence on a third variable, current velocity. Support for this view comes from the experimental work of Harms (1969, p.379) wherein he demonstrated that within the size range of 0.2 to 0.5 mm grain size exerts little influence on the occurrence of sedimentary structures (in contrast to flow velocity). Thus, for all sands within this size range a given current will produce a given structure. However, outside of this limited range other structures are produced by the same current. This latter effect is represented in our studies by the occurrence of lag ripple marks. This structure tends to occur with streaming lineation on the same surface. The streaming lineation is developed in larger but less dense sand grains whereas lag ripples are composed of sand-size heavy minerals. This illustration provides an example of the hydrodynamic separation of sediment according to density. A given current can simultaneously produce two different sedimentary structures.

TABLE XIII

Percent mud in sedimentary structures

| *Sedimentary structure* | *Weight percent mud* *average* | *range* | *No. of measurements* |
|---|---|---|---|
| Streaming lineation | 11.3 | 1.8–47.2 | 9 |
| Cuspate ripples | 13.0 | 2.0–45.1 | 13 |
| Sinuous ripples | 20.0 | 3.1–83.5 | 12 |
| Shrinkage crack | 34.8 | 13.6–78.2 | 6 |
| Parting lineation | 51.2 | 11.6–70.9 | 6 |
| Linear ripples | 54.0 | 14.7–88.4 | 5 |

Subenvironment is a factor that controls the distribution of sedimentary structures by operating to select the processes of erosion and deposition that occur at that point. At different sites within the fluvial environment different sedimentary processes operate to produce different suites of sedimentary features.

Flow velocity is the most important factor in directly producing the wide variety of common sedimentary features. With the exception of post-depositional structures, most of the other types of structures and bedding are primarily controlled by this parameter. Flow velocity affects the sediment in several ways, including: the type and extent of turbulence, balance between erosion and deposition, and mode of transportation.

The significance of turbulence can be seen in the differently shaped scours produced by helical flow and current vortices. Although the presence of channel obstructions is usually necessary to produce these features, the magnitude and intensity of scouring is directly proportional to the flow velocity.

The effect of water velocity on the balance between erosion and deposition is illustrated by Hjulström's diagram (1935, p.298), which relates fields of erosion, transportation and deposition to the parameters of velocity and particle size. In a stream where particle size is an independent variable determined by climate and lithology in the source area, flow velocity is the critical variable in determining whether the sediment is eroded or deposited. In ephemeral streams that are characterized by variable episodic flow, the influence of flow velocity is magnified. Not only does flow velocity change progressively during a flood, giving rise to a predictable sequence of sedimentary features, but also the magnitude varies from flood to flood. Thus, the deposits of a small flood have different sedimentary characteristics than those of a large flood. This feature itself is a useful criterion for recognition of ancient ephemeral streams.

The overriding importance of flow velocity is best seen among sedimentary structures formed during transportation and deposition. As the flow velocity increases, an orderly succession of sedimentary structures and associated bedding types is produced. This succession from lowest flow velocity to highest is: parting lineation (horizontal parallel stratification), linquoid ripples (micro cross-stratification), cuspate ripples (micro cross-stratification), sinuous ripples (micro cross-stratification), lag ripples (horizontal discontinuous stratification), streaming lineation (horizontal discontinuous stratification), and antidunes (backset cross-stratification).

## STRUCTURES AND STRATIFICATION

The utility of sedimentary structures for interpreting ancient environments of deposition has been amply demonstrated in many studies. Interpretations based upon only a small number of structures are generally less reliable than interpretations based on larger numbers of structures. This problem is particularly acute in fluvial formations, which are known for their scarcity of sedimentary structures. However, nearly all sedimentary formations contain abundant stratification. Unfortunately, the origin of various types of stratification is not as well understood as the origins of sedimentary structures, thereby decreasing the interpretative value of the most abundant sedimentary features (stratification). One of the goals of this book has been to relate sedimentary structures to specific stratification types so that more precise and reliable interpretations can be made in the absence of sedimentary structures. The results are listed in Table XIV, which shows stratification types associated with or produced by various sedimentary structures. Of particular significance is the association of various types of horizontal and cross-stratification with corresponding flow-regime interpretations.

TABLE XIV

Stratification and associated features

| *Stratification type* | *Associated features* | *Interpretation* |
|---|---|---|
| Horizontal parallel | parting lineation | low flow velocity |
| Horizontal discontinuous | streaming lineation | upper-flow regime |
| Lenticular | | shifting depositional centers, or randomly fluctuating conditions |
| Graded | | waning currents or suspension deposit |
| Micro cross-stratification | cuspate ripples | lower-flow regime |
| Festoon cross-stratification | dunes | lower-flow regime |
| Ripple | linear ripples | lower-flow regime |
| Inclined | horizontal discontinuous | upper-flow regime |
| Low-angle cross-stratification | horizontal discontinuous | upper-flow regime |
| Avalanche-front cross-stratification | | suspension deposit, or lower-flow regime |
| Backset cross-stratification | antidunes | upper-flow regime |
| Disturbed | convolute stratification, flame structure, flow roll | slump |
| Disturbed | burrows, mottled | bioturbation |
| Disturbed | intraformational conglom., shrinkage crack | desiccation |
| Disturbed | stepped profile | erosion surface |

## PRESERVATION

The occurrence of sedimentary structures is obviously dependent as much on preservation as upon conditions of formation. Unfortunately, our knowledge of the preservation potential of different types of sedimentary structures lags far behind our knowledge of the origin of those structures. Although our study adds little in terms of quantitative information to this problem, it does tend to confirm many common assumptions about modes of preservation. First, the preservation of bedding-plane structures is less likely than the preservation of their associated bedding types. For example, micro cross-stratification is a common bedding type found in many sedimentary formations, in contrast with cuspate ripple marks which are relatively rare. Secondly, those structures that project downward into the body of the sediment are more likely to be preserved than are bedding plane features. The various types of disturbed stratification are examples of the first category and ripple marks are representative of the second. Thirdly, features that occur near the tops of sedimentation units are more likely to be destroyed by subsequent events than are features that occur lower in sedimentation units. Examples include shrinkage cracks and streaming lineation, respectively. Fourthly, point-bar and flood-plain deposits are more favorably situated for preservation than are channel deposits. The latter are likely to be scoured by subsequent floods but the former will tend to be preserved. Thus, linear ripple marks and parting lineation are more readily preserved than cuspate ripple marks and streaming lineation. Finally, since the surface of terrestrial deposits is exposed to reworking by water, wind, and organisms, many features in the upper few inches of sedimentation units are readily destroyed. However, features that are produced by these elements may be preserved because they are in equilibrium with atmospheric conditions. Thus textured surfaces probably are a common bedding plane feature in many terrestrial deposits. If textured surfaces were more striking, they would probably be recognized as one of the most abundant structures in terrestrial rocks.

## SEDIMENTARY STRUCTURES IN FLUVIAL DEPOSITS

Fluvial rocks contain a limited suite of sedimentary structures. In general, several types of horizontal and cross-stratification are common, but most sedimentary structures are rare. Among sedimentary structures the most abundant types are erosional and post-depositional structures. Structures formed during transportation and deposition are poorly represented; cuspate ripple marks, parting lineation and flute casts are the most abundant. This unusual distribution of sedimentary features can be understood if consideration is given to both modes of formation and preservation. Scour structures tend to be preserved because they are rapidly buried by deposits during the waning stages of floods. Similarly, post-depositional structures are present in abundance in fluvial rocks because they tend to form after burial (slump structures) or within the body of sediment (bioturbation features). The great abundance of horizontal and cross-stratification in fluvial rocks and the scarcity of ripple marks is readily explained by the mechanics of stream deposition. Under conditions of episodic flow the channel is first eroded and then backfilled as the flood passes. In ephemeral streams the bulk of deposition takes place in upper-flow regime conditions, resulting in thick units of inclined or low-angle cross-stratification characterized by horizontal discontinuous stratification. Lower-flow regime conditions are rather short-lived and only thin veneers of ripple-marked deposits are formed at the tops of sedimentation units. These deposits are rarely preserved because subsequent floods scour the bottom, leaving behind only the lower portions of previous sedimentation units. Eventually the channel is filled with deposits characterized by horizontal discontinuous stratification and lower-flow regime bed forms are scarce. Only those sedimentary structures that originally were produced in great abundance can be preserved as common sedimentary features in fluvial rocks. Cuspate ripple marks are the only structure formed during transportation and deposition that is in great abundance along ephemeral streams. Thus, although many are destroyed, enough cuspate ripples remain to be present in rare to common amounts in fluvial rocks. The relatively high abundance of parting lineation observed in fluvial rocks cannot be readily explained, either in terms of its abundance in modern streams, or its position at the top of sedimentation units. However, parting lineation frequently has been confused with streaming lineation, a structure that is both more abundant and more favorably positioned for preservation, and the high estimates of the abundance of parting lineation in ancient rocks may not be real.

## ASSOCIATION OF SEDIMENTARY STRUCTURES

In any depositional environment sedimentary features should form natural groups representing the products of similar processes of formation, sediment types, subenvironments and so forth. These groupings should be especially pronounced in fluvial regimes where processes markedly differ within small areas. To test this hypothesis, the information in Table II was quantified by assigning numerical values to each abundance class (abundant = *4*, common = *3*, rare = *2*, trace = *1*, not observed = *0*). The data were analyzed by factor analysis (using the Biomedical package from Stanford University) in order to determine the composition of natural groupings within the data. Each resulting factor (Table XV) represents a naturally occurring group of sedimentary structures. Within each group the sedimentary structures are more closely related to each other than they are to other groups or structures listed in other groups. Consideration of the composition of each group generally indicates the factors of the depositional environment responsible for the grouping. Group 1 represents low-to-moderate-velocity channel deposits including thalweg and longitudinal and transverse bars. Group 2 is less well-defined than Group 1, but basically represents channel-margin deposits, including lateral and point bars, which have been affected by lower-flow regime currents and post-depositional disturbance. Group 3 is also a compound grouping representing channel margins, deposits similar to those of Group 2, but showing the effects of primary currents, saline features or quicksand disturbance. Group 4 represents post-depositional modifications of channel-margin deposits and banks. The occurrence of imbrication in this group is anomalous and this structure is only tenuously connected to this group. Group 5 records post-depositional surface markings that are present in all subenvironments. Group 6 has no environmental meaning. Group 7 represents floodplain deposits. Group 8 represents high-velocity channel deposits. Group 9 is meaningless. Group 10 represents scour marks on channel-bar and channel-margin deposits. Although this approach is not without ambiguity, the groupings are significant and provide a further insight into the mechanics of sedimentary processes along ephemeral streams.

## ENVIRONMENTAL CRITERIA

Several general observations concerning the utility of sedimentary structures for reconstructing environments of deposition can be drawn from the information presented in this book. First, there is no single structure or group of structures uniquely indicative of ephemeral streams. Although several features, such as micro-terraces, dendritic marks, and horizontal discontinuous stratification, are strongly suggestive of episodic flow in ephemeral streams, none are definitive in themselves. Secondly, among all environments ephemeral streams most closely resemble tidal flats and channels. Both environments are characterized by episodic flow and waxing and waning currents. Despite the close resemblance of their sedimentary features, the great difference in regional geologic setting should prevent confusion of deposits of these two environments. Thirdly, among fluvial environments ephemeral streams more closely resemble braided streams than meandering streams. Flow in meandering streams is rather uniform and thick deposits characterized by ripple or micro cross-stratification may occur. In contrast, flow in braided streams is episodic, although not to the extent as in ephemeral streams. Fourthly, because different processes operate at different sites within the fluvial complex, subenvironments are readily recognized on the basis of their characteristic sedimentary features. Fifthly, sedimentation units are even more useful than sedimentary features in distinguishing subenvironments. In

TABLE XV

Associations of sedimentary structures

| *1* | *2* | *3* | *4* | *5* | *6* | *7* | *8* | *9* | *10* |
|---|---|---|---|---|---|---|---|---|---|
| rib-and-furrow | scour hole | armored mud ball | micro-terrace | cracks | dendritic mark | earth crack | crescent scour | rill | tool mark |
| cuspate ripple | secondary ripple | parting lineation | fluted step | raindrop impression | algal mat | algal mound | lag ripple | textured surface | groove |
| sinuous ripple | longitudinal ripple | mud pebble | linear ripple | | | | streaming lineation | | |
| eolian ripple | polygonal shrinkage crack | reverse mud curl<br>salt crust | imbrication | | | | | | |
| chevron ripple | linear shrinkage crack | salt ridge | | | | | | | |
| rhomboid ripple | | wrinkled surface | | | | | | | |
| cuspate and secondary ripple | multiple secondary ripple | crystal mold | | | | | | | |
| linear and secondary ripple | mud curl | gas bubble<br>sand volcanoe | | | | | | | |

TABLE XVI

Directional properties of sedimentary structures

| *Structure* | *No. of meas-urements* | *Angle about channel* | | *Angle from channel* | |
|---|---|---|---|---|---|
| | | *mean* | *std. dev.* | *mean* | *std. dev.* |
| Unimodal patterns | | | | | |
| parting lineation | 88 | 3° | 17° | 11° | 14° |
| cuspate ripples | 818 | 0° | 18° | 13° | 14° |
| sinuous ripples | 237 | 2° | 17° | 13° | 12° |
| streaming lineation | 560 | 1° | 16° | 12° | 11° |
| crescent scours | 74 | 1° | 11° | 8° | 8° |
| longitudinal ripples | 106 | 5° | 16° | 12° | 11° |
| Bimodal patterns | | | | | |
| linear ripples | 113 | 6° | 69° | 61° | 33° |
| rills | 35 | 10° | 76° | 70° | 30° |
| secondary ripples | 10 | 11° | 83° | 74° | 30° |
| low-angle cross-strat. | 42 | 8° | 79° | 67° | 41° |
| avalanche cross-strat. | 27 | 25° | 64° | 55° | 40° |
| Miscellaneous (too few measurements to determine pattern) | | | | | |
| imbrication | 7 | 3° | 14° | 11° | 8° |
| grooves | 5 | 3° | 3° | 3° | 3° |
| festoon cross-strat. | 4 | 48° | 70° | 60° | 56° |

addition to containing most of the characteristic sedimentary features, each autocyclic sequence also records depositional history. These sedimentation units are the most reliable criteria for differentiating various fluvial subenvironments.

## PALEOCURRENT INDICATORS

Many of the sedimentary structures described here are potentially useful as indicators of current direction. To be useful paleocurrent indicators, structures should be asymmetric in shape and preferentially aligned relative to the current. Alignment of various structures was tested by measuring the structure direction and the channel direction at more than 2000 sites along ephemeral streams. Table XVI lists the results of this study in order of decreasing reliability as current indicators. Note the relatively low ranking of both low-angle and avalanche-front cross-stratification, two structures that are the basis for the majority of paleocurrent studies. Our data indicates that these structures should be used only with caution. Both of these features tend to show bimodal distributions about the channel direction. Although the average of all directions or the bisector of the two separate modes corresponds closely with the channel direction, large numbers of measurements are required to define the bimodal pattern. With smaller numbers of measurements sampling error is a problem and only one mode is defined, thereby biasing the reconstructed current direction. Few outcrops contain sufficient numbers of cross-stratification sets to permit the use of low-angle or avalanche-front cross-stratification as reliable current indicators.

An associated problem concerns the degree of scatter of paleocurrent measurements in different environments. Commonly it is assumed that fluvial paleocurrent patterns show markedly less scatter than patterns developed in shallow-marine or lacustrine environments. In actuality, however, the degree of scatter between fluvial and marine environments is nearly equal. Potter and Pettijohn (1963, p.89) report variances of 4000–6000 and 6000–8000 for fluvial and marine paleocurrent studies, respectively. However, in view of the data presented here (Table XVI), it is probable that fluvial environments do have a lesser degree of scatter than marine environments. The inclusion of structures that are bimodaly distributed about the main current direction, such as low-angle cross-stratification, have greatly increased the apparent scatter of resulting paleocurrent patterns. In contrast, the variance for those structures, such as cuspate ripples, that are oriented downcurrent is significantly lower and probably represents a true representation of the variance for fluvial paleocurrents (Table XVI).

## PROXIMAL-DISTAL RELATIONS

Changes in conditions and features along the course of small streams are generally ignored. In most studies of both ancient and modern streams, conditions are considered to be basically uniform along the length of the channel. Although obvious differences between upstream and downstream reaches are acknowledged, more subtle variations pass unrecognized. However, changes do occur, some of which are systematic and progressive. The recognition of these gradients would aid in detailed reconstructions of fluvial environments. Downstream changes that were noted during the course of this study include: increase in channel sinuosity, decrease in grain size, and decrease in flow velocity as determined by sedimentary structures.

Stream sinuosity was determined by measuring both channel and structure directions at 50 intervals at several sites along the streams. Although the structures tend to have slightly more scatter than the channel, both show consistent trends. In general, the ephemeral streams studied tend to be straight to sinuous through most of their length, becoming meandering in lower reaches.

Grain size also decreases downstream, largely as a result of the progressive loss of coarse particles. In contrast, the size of individual gravel particles tends to remain constant even though the proportion of gravel in the sediment decreases. This observation (based on the measurement of 969 pebbles in five sites along a 21-miles stretch of Ashley Creek) is strongly suggestive of transportation by suspension rather than traction. This mode of transportation is indicated despite the large size of the average gravel particle (mean weight of 969 pebbles is 6.1 kg) indicating the magnitude of the infrequent episodic events and their potential for accomplishing geologic work. This observation agrees with inferences about sediment transportation and deposition drawn from the great abundance of horizontal discontinuous stratification.

The final downstream trend observed relates to the distribution of structures formed during transportation and deposition. Upstream reaches are characterized by streaming lineation, crescent scours and sinuous ripples. Downstream reaches are characterized by cuspate ripples, parting lineation and shrinkage cracks. The inferred downstream decrease in flow velocity is confirmed by observation of flows in ephemeral streams in which water is progressively lost downstream because of evaporation and infiltration.

We conclude this book by again stressing that ephemeral streams form a middle ground between flume experiments and field studies of sedimentary rocks. Because we are geologists, who through most of our careers have been primarily interested in rocks, we hope that we have contributed to the study of fluvial sedimentary rocks.

## REFERENCES


Allen, J. R. L., 1961. Sandstone-plugged pipes in the Lower Old Red Sandstone of Shropshire, England. *J. Sediment. Petrol.*, 31: 325–335.

Allen, J. R. L., 1962. Petrology, origin, and deposition of the highest Lower Old Red Sandstone of Shropshire, England. *J. Sediment. Petrol.*, 32: 657–697.

Allen, J. R. L., 1963. The classification of cross-stratified units with notes on their origin. *Sedimentology*, 2: 93–114.

Allen, J. R. L., 1964. Primary current lineation in the lower Old Red Sandstone (Devonian), Anglo-Welsh Basin. *Sedimentology*, 3: 89–108.

Allen, J. R. L., 1965a. Scour marks in snow. *J. Sediment. Petrol.*, 35: 331–338.

Allen, J. R. L., 1965b. Fining-upward cycles in alluvial successions. *Geol. J.*, 4: 229–246.

Allen, J. R. L., 1965c. A review of the origin and characteristics of recent alluvial sediments. *Sedimentology*, 5: 89–191.

Allen, J. R. L., 1968. *Current Ripples.* North-Holland, Amsterdam, 433 pp.

Allen, J. R. L., 1969. Erosional current marks of weakly cohesive mud beds. *J. Sediment. Petrol.*, 31: 607–623.

Allen, J. R. L., 1970a. Studies in fluviatile sedimentation: a comparison of fining-upward cyclothems, with special reference to coarse-member composition and interpretation. *J. Sediment. Petrol.*, 40: 293–323.

Allen, J. R. L., 1970b. A quantitative model of climbing ripples and their cross-laminated deposits. *Sedimentology*, 14: 5–26.

Allen, J. R. L., 1970c. A quantitative model of grain size and sedimentary structures in lateral deposits. *Geol. J.*, 7: 129–146.

Anderson, J. J. and Everett, J. R., 1964. Mudcrack formation studied by time-lapse photography. *Geol. Soc. Am., Progr. Ann. Meeting.* 4–5.

Bagnold, R. A., 1941. *The Physics of Blown Sand and Desert Dunes.* Methuen, London, 265 pp.

Barnes, W. C. and Smith, A. G., 1964. Some markings associated with ripplemarks from the Proterozoic of North America. *Nature*, 201: 1018–1019.

Barrell, J., 1913. The Upper Devonian delta of the Appalachian geosyncline, I. The delta and its relations to the interior sea. *Am. J. Sci.*, 36: 429–472.

Bell, H. S., 1940. Armored mud balls: their origin, properties and role in sedimentation. *J. Geol.*, 48: 1–31.

Belt, E. S., 1968. Carboniferous continental sedimentation, Atlantic Provinces, Canada. In: G. de V. Klein, (Editor), *Late Paleozoic and Mesozoic Continental Sedimentation, Northeastern North America–Geol. Soc. Am., Spec. Pap.*, 106: 127–176.

Bernard, H. A., Major, C. F., Jr., Parrott, B. S. and LeBlanc, R. J., Sr., 1970. Recent sediments of southeast Texas. *Bur. Econ. Geol., Univ. Texas, Austin, Guidebook*, 11.

Bluck, B. J., 1967. Deposition of some Upper Old Red Sandstone conglomerate in the Clyde area: a study in the significance of bedding. *Scot. J. Geol.*, 3: 139–167.

Boswell, P. G. H., 1961. *Muddy Sediments.* W. Heffer, Cambridge, 140 pp.

Boyd, D. W. and Ore, H. T., 1963. Patterned cones in Permo-Triassic red beds of Wyoming and adjacent areas. *J. Sediment. Petrol.*, 33: 438–451.

Bradley, W. H., 1929. Algae reefs and oolites of the Green River Formation. *U.S. Geol. Surv., Profess. Pap.*, 154: 203–223.

Bradley, W. H., 1933. Factors that determine the curvature of mudcracked layers. *Am. J. Sci.*, 26: 55–71.

Brush, L. M., Jr., 1965. Experimental work on primary sedimentary structures. In: G. V. Middleton (Editor), *Primary Sedimentary Structures and their Hydrodynamic Interpretation–Soc. Econ. Paleontologists Mineralogists, Spec. Publ.*, 12: 17–24.

Bucher, W. H., 1919. On ripples and related sedimentary surface forms and their paleogeographic interpretation. *Am. J. Sci.*, 47: 149–210; 241–267.

Burst, J. F., 1965. Subaqueously formed shrinkage cracks in clay. *J. Sediment. Petrol.*, 35: 348–353.

Cailleux, A., 1945. Distinction des galets marins et fluviatiles. *Bull. Soc. Géol. France*, 15: 375–404.

Carozzi, A. V., 1962. Observations on algal bioherms in the Great Salt Lake, Utah. *J. Geol.*, 70: 246–253.

Choquette, P. W. and Traut, J. D., 1963. Pennsylvanian carbonate reservoirs, Ismay field, Utah and Colorado. In: *Shelf Carbonates of the Paradox Basin–Four Corners Geol. Soc., Field Conf., 4th*, pp.157–184.

Cloud, P. E., Jr., 1960. Gas as a sedimentary and diagenetic agent. *Am. J. Sci.*, 258-A: 35–45.

Cloud, P. E., Jr., 1968. Pre-metazoan evolution and the origins of the metazoa. In: E. T. Drake (Editor), *Evolution and Environment.* Yale Univ. Press, New Haven–London, pp.1–72.

Collinson, J. D., 1970. Bedforms of the Tana River, Norway. *Geogr. Annaler.*, 52-A: 31–56.

Conybeare, C. E. B. and Crook, K. A. W., 1968. *Manual of Sedimentary Structures–Bull. Bur. Min. Resources*, 102: 327 pp.

Cooper, J. R., 1943. Flow structures in the Berea Sandstone and Bedford Shale of central Ohio. *J. Geol.*, 51: 190–203.

Culbertson, J. K. and Scott, C. H., 1970. Sand-bar development and movement in an alluvial channel, Rio Grande near Bernardo, New Mexico. *U.S. Geol. Surv., Profess. Pap.*, 700B: 237–241.

Davies, H. G., 1965. Convolute lamination and other structures from the Lower Coal Measures of Yorkshire. *Sedimentology*, 5: 305–325.

Dawson, J. W., 1890. On burrows and tracks of invertebrate animals in Paleozoic rocks and other markings. *Q. J. Geol. Soc. Lond.*, 46: 595–618.

Dickas, A. B. and Lunking, W., 1968. The origin and destruction of armored mud balls in a fresh-water lacustrine environment. *J. Sediment. Petrol.*, 38: 1366–1370.

Dmitrieva, E. V., Erchova, G. I. and Orechnikova, E. I., 1962. *Structures and Textures of Sedimentary Rocks, 1. Clastic and Argillaceous Rocks.* Gosgeoltekhizdat, Moscow (in Russian with French summary).

Doeglas, D. J., 1962. The structure of sedimentary deposits of braided streams. *Sedimentology*, 1: 167–190.

Dorr, J. A., Jr. and Kauffman, E. G., 1963. Rippled toroids from the Napoleon Sandstone Member (Mississippian) of southern Michigan. *J. Sediment. Petrol.*, 33: 751–758.

Dow, W. G., 1964. The effect of salinity on the formation of mud-cracks. *Compass*, 41: 162–166.

Dzulynski, S. and Sanders, J. E., 1962. Current marks on firm mud bottoms. *Trans. Conn. Acad. Arts Sci.*, 42: 57–96.

Dzulynski, S. and Walton, E. K., 1965. *Sedimentary Features of Flysch and Greywackes.* Elsevier, Amsterdam, 274 pp.

Emery, K. O., 1945. Entrapment of air in beach sand. *J. Sediment. Petrol.*, 15: 39–49.

Emery, K. O., 1950. Contorted Pleistocene strata at Newport Beach, California. *J. Sediment. Petrol.*, 20: 111–115.

Fahnestock, R. K., 1963. Morphology and hydrology of a glacial stream, White River, Mount Rainier, Washington. *U.S. Geol. Surv., Profess. Pap.*, 422-A: 1–70.

Fairbridge, R. W., 1947. Possible causes of intraformational disturbances in the Carboniferous varve rocks of Australia. *R. Soc. N.S.W.*, 81: 99–121.

Folk, R. L., 1971. Longitudinal dunes of the northwestern edge of the Simpson desert, Northern Territory, Australia, 1. Geomorphology and grain-size relationships. *Sedimentology*, 16: 5–54.

Folk, R. L. and Ward, W. C., 1957. Brazos River bar: a study in the significance of grain-size parameters. *J. Sediment. Petrol.*, 27: 3–26.

Frarey, M. J. and McLaren, D. J., 1963. Possible metazoans from the Early Proterozoic of the Canadian Shield. *Nature*, 200: 461–462.

Frarey, M. J., Ginsburg, R. N. and McLaren, D. J., 1963. Metazoan tubes from the type Huronian, Ontario, Canada. *Geol. Soc. Am., Abstr.*, p.63A.

Fraser, H. J., 1935. Experimental study of the porosity and permeability of clastic sediments. *J. Geol.*, 43: 910–1010.

Frazier, D. E. and Osanik, A., 1961. Point-bar deposits, Old River Locksite, Louisiana. *Trans. Gulf Coast Assoc. Geol. Soc.*, 11: 121–137.

Friend, P. F., 1965. Fluvial sedimentary structures in the Wood Bay Series (Devonian) of Spitzbergen. *Sedimentology*, 5: 39–68.

Gill, W. D. and Kuenen, P. H., 1958. Sand volcanoes on slumps in the Carboniferous of County Clare, Ireland. *Q. J. Geol. Soc. Lond.*, 113: 441–460.

Grant-Mackie, J. A. and Lowrey, D. C., 1964. Upper Triassic rocks of Kiritehere, southwest Auckland, New Zealand. *Sedimentology*, 3: 296–317.

Grim, R. E. and Allen, V. T., 1938. Petrology of the Pennsylvanian underclays of Illinois. *Bull. Geol. Soc. Am.*, 41: 1485–1513.

Gubler, J., Bugnicourt, D., Faber, J., Kurler, B. and Nyssen, R., 1966. *Essai de Nomenclature et Caractérisation des principales Structures sedimentaires.* Editions Technip, Paris, 291 pp.

Hamblin, W. K., 1961. Micro cross-lamination in Upper Keweenawan sediments of northern Michigan. *J. Sediment. Petrol.*, 31: 390–401.

Hand, B. M., 1969. Antidunes as trochoidal waves. *J. Sediment. Petrol.*, 39: 1302–1309.

Hand, B. M., Wessell, J. M. and Hayes, M. O., 1969. Antidunes in the Mount Toby Conglomerate (Triassic), Massachusetts. *J. Sediment. Petrol.*, 39: 1310–1316.

Häntzschel, W., 1938. Bau und Bildung von Grob-Rippeln im Wattenmeer. *Senckenbergiana*, 20: 1–42.

Häntzschel, W., 1949. Zur Deutung von *Manchuriophycus endo* und ähnlichen Problematika. *Geol. Staatsinst. Hamburg*, 19: 77–84.

Häntzschel, W., 1962. Trace fossils and problematica. In: R. C. Moore (Editor), *Treatise on Invertebrate Paleontology, Part W–Geol. Soc. Am.*, pp.W177–W245.

Harms, J. C., 1966. Stratigraphic traps in a valley fill, western Nebraska. *Bull. Am. Assoc. Petrol. Geologists*, 50: 2119–2149.

Harms, J. C., 1969. Hydraulic significance of some sand ripples. *Bull. Geol. Soc. Am.*, 80: 363–396.

Harms, J. C. and Fahnestock, R. K., 1965. Stratification, bed forms, and flow phenomena (with example from the Rio Grande). *Soc. Econ. Paleontologists Mineralogists, Spec. Publ.*, 12: 84–115.

Harms, J. C., Mackenzie, D. B. and McCubbin, D. G., 1963. Stratification in modern sands of the Red River, Louisiana. *J. Geol.*, 71: 566–580.

High, L. R., Jr. and Picard, M. D., 1968. Dendritic surge marks *(Dendrophycus)* along modern stream banks. *Contrib. Geol.*, 7: 1–6.

High, L. R., Jr., Hepp, D. M., Clark, T. and Picard, M. D., 1969. Stratigraphy of Popo Agie Formation (Late Triassic), Uinta Mountain area, Utah and Colorado. In: *Geologic Guidebook of the Uinta Mountains–Intermountain Assoc. Geol., Field Conf., 16th:* 181–192.

Hjulström, F., 1935. Studies of the morphological activity of rivers as illustrated by the River Fyris. *Bull. Geol. Inst. Univ. Uppsala*, 25: 227–346.

Hoyt, J. H. and Henry, V. J., Jr., 1963. Rhomboid ripple mark, indicator of current direction and movement. *J. Sediment. Petrol.*, 33: 604–608.

Hunter, R. E., 1969. Eolian microridges on modern beaches and a possible ancient example. *J. Sediment. Petrol.*, 39: 1573–1578.

Imbrie, J. and Buchanan, H., 1965. Sedimentary structures in modern carbonate sands of the Bahamas. *Soc. Econ. Paleontologists Mineralogists, Spec. Publ.*, 12: 149–172.

Jopling, A. V., 1960. *An Experimental Study on the Mechanics of Bedding.* Thesis, Harvard Univ., Cambridge, Mass., 358 pp.

Jopling, A. V., 1961. Origin of regressive ripples explained in terms of fluid-mechanic processes. *U.S. Geol. Surv., Profess. Pap.*, 424: 15–17.

Jopling, A. V., 1963. Hydraulic studies on the origin of bedding. *Sedimentology*, 2: 115–121.

Jopling, A. V., 1964. Interpreting the concept of the sedimentation unit. *J. Sediment. Petrol.*, 34: 165–172.

Jopling, A. V., 1966. Some applications of theory and experiment to the study of bedding genesis. *Sedimentology*, 7: 71–102.

Jopling, A. V. and Richardson, E. V., 1966. Backset bedding developed in shooting flow in laboratory experiments. *J. Sediment. Petrol.*, 36: 821–825.

Jopling, A. V. and Walker, R. G., 1968. Morphology and origin of ripple-drift cross-lamination, with examples from the Pleistocene of Massachusetts. *J. Sediment. Petrol.*, 38: 971–984.

Jüngst, H., 1934. Geological significance of synaeresis. *Geol. Rundsch.*, 25: 321–325.

Karcz, I., 1966. Secondary currents and the configuation of a natural stream bed. *J. Geophys. Res.*, 71: 3109–3112.

Karcz, I., 1967. Harrow marks, current-aligned sedimentary structures. *J. Geol.*, 75: 113–121.

Karcz, I., 1968. Fluviatile obstacle marks from the wadis of the Negev (southern Israel). *J. Sediment. Petrol.*, 38: 1000–1012.

Karcz, I., 1969. Mud pebbles in a flash floods environment. *J. Sediment. Petrol.*, 39: 333–337.

Karcz, I., and Goldberg, M., 1967. Ripple-controlled desiccation patterns from Wadi Shiqma, southern Israel. *J. Sediment. Petrol.*, 37: 1244–1245.

Kelling, G. and Williams, B. P. J., 1966. Deformation structures of sedimentary origin in the Lower Limestone Shales (basal Carboniferous) of south Pembrokeshire, Wales. *J. Sediment. Petrol.*, 36: 927–939.

Kendall, C. G. St. C. and Skipwith, P. A. d'E., 1969. Holocene shallow-water carbonate and evaporite sediments of Khoral Bazam, Abu Dhabi, southwest Persian Gulf. *Bull. Am. Assoc. Petrol. Geologists*, 53: 841–869.

Kindle, E. M., 1916. Small pit and mound structures developed during sedimentation. *Geol. Mag.*, 3: 542–547.

Kindle, E. M., 1917. Recent and fossil ripple mark. *Mus. Bull., Geol. Surv. Can.*, 25: 1–56.

Klein, G. de V., 1965. Dynamic significance of primary structures in the Middle Jurassic Great Oolite Series, southern England. *Soc. Econ. Paleontologists Mineralogists, Spec. Publ.*, 12: 173–191.

Klein, G. de V., 1970. Depositional and dispersal dynamics of intertidal sand bars. *J. Sediment. Petrol.*, 40: 1095–1127.

Krumbein, W. C., 1939. Preferred orientation of pebbles in sedimentary deposits. *J. Geol.*, 47: 673–706.

Krumbein, W. C., 1940. Flood gravel of San Gabriel Canyon, California. *Bull. Geol. Soc. Am.*, 51: 639–676.

Krumbein, W. C., 1942. Flood deposits of Arroyo Seco, Los Angeles County, California. *Bull. Geol. Soc. Am.*, 53: 1355–1402.

Lachenbruch, A. H., 1961. Depth and spacing of tension cracks. *J. Geophys. Res.*, 66: 4273–4292.

Lane, E. W. and Carlson, E. J., 1954. Some observations of the effect of particle shape on movement of coarse sediments. *Trans. Am. Geophys. Union*, 35: 453–462.

Laporte, L. F., 1968. *Ancient Environments*. Prentice-Hall, Englewood Cliffs, N.J., 116 pp.
Leighley, J. B., 1934. Turbulence and the transportation of rock debris by streams. *Geograph. Rev.*, 24: 453–464.
Leney, G. W. and Leney, A. T., 1957. Armored fill balls in the Pleistocene outwash of southeastern Michigan. *J. Geol.*, 65: 105–106.
Leopold, L. B. and Maddock, T., Jr., 1953. The hydraulic geometry of stream channels and some physiographic implications. *U.S. Geol. Surv., Profess. Pap.*, 252: 57 pp.
Leopold, L. B. and Miller, J. P., 1956. Ephemeral streams–hydraulic factors and their relation to the drainage net. *U.S. Geol. Surv., Profess. Pap.*, 282-A: 1–37.
Leopold, L. B. and Wolman, M. G., 1957. River channel patterns: braided, meandering and straight. *U.S. Geol. Surv., Profess. Pap.*, 282-B: 39–85.
Leopold, L. B., Wolman, M. G. and Miller, J. P., 1964. *Fluvial Processes in Geomorphology*. Freeman, San Francisco, Calif., 522 pp.
Longwell, C. R., 1928. Three common types of desert mud cracks. *Am. J. Sci.*, 15: 136–145.
Lowry, W. D. and Cooper, B. N., 1970. Penecontemporaneous downdip slump structures in Middle Ordovician limestone, Harrisonburg, Virginia. *Bull. Am. Assoc. Petrol. Geologists*, 54: 1938–1945.
Lyell, C., 1851. On fossil rain marks of the recent, Triassic and Carboniferous Periods. *Q. J. Geol. Soc. Lond.*, 7: 238–247.
Matthes, G., 1947. Macroturbulence in natural streams. *Trans. Am. Geophys. Union*, 28: 255–261.
Maxon, J. H., 1940. Gas pits in non-marine sediments. *J. Sediment. Petrol.*, 10: 142–145.
McBride, E. F. and Yeakel, L. S., 1963. Relationship between parting lineation and rock fabric. *J. Sediment. Petrol.*, 33: 779–782.
McCave, I. N., 1969. Correlation of marine and non-marine strata with example from Devonian of New York State. *Bull. Am. Assoc. Petrol. Geologists*, 53: 155–162.
McCormick, C. D. and Picard, M. D., 1969. Petrology of Gartra Formation (Triassic), Uinta Mountain area, Utah and Colorado. *J. Sediment. Petrol.*, 39: 1484–1508.
McGowen, J. H. and Garner, L. E., 1970. Physiographic features and stratification types of coarse-grained point bars: modern and ancient examples. *Sedimentology.*, 14: 77–111.
McKee, E. D., 1938. Original structures in Colorado River flood deposits of Grand Canyon. *J. Sediment. Petrol.*, 8: 77–83.
McKee, E. D., 1945. Small-scale structures in the Coconino Sandstone of northern Arizona. *J. Geol.*, 53: 313–325.
McKee, E. D., 1954. Stratigraphy and history of the Moenkopi Formation of Triassic age. *Geol. Soc. Am., Mem.*, 61: 133 pp.
McKee, E. D., 1957. Flume experiments on the production of stratification and cross-stratification. *J. Sediment. Petrol.*, 27: 129–134.
McKee, E. D., 1964. Inorganic sedimentary structures. In: *Approaches to Paleoecology*. Wiley, New York, N.Y., pp.275–295.
McKee, E. D., 1965. Experiments on ripple lamination. *Soc. Econ. Paleontologists Mineralogists, Spec. Publ.*, 12: 66–83.
McKee, E. D., 1966. Significance of climbing-ripple structure. *U.S. Geol. Surv., Profess. Pap.*, 550-D: 94–103.
McKee, E. D. and Weir, G. W., 1953. Terminology for stratification and cross-stratification in sedimentary rocks. *Bull. Geol. Soc. Am.*, 64: 381–390.
McKee, E. D., Crosby, E. J. and Berryhill, H. L., Jr., 1967. Flood deposits, Bijou Creek, Colorado, June 1965. *J. Sediment. Petrol.*, 37: 829–851.
McKee, E. D., Reynolds, M. A. and Baker, C. H., Jr., 1962a. Laboratory studies on deformation in unconsolidated sediment. *U.S. Geol. Surv., Profess. Pap.*, 450D: 151–155.

McKee, E. D., Reynolds, M. A. and Baker, C. H., Jr., 1962b. Experiments on intraformational recumbant folds in crossbedded sand. *U.S. Geol. Surv., Profess. Pap.*, 450D: 155–160.

Middleton, G. V., 1965. Antidune cross-bedding in a large flume. *J. Sediment. Petrol.*, 35: 922–927.

Middleton, G. V., 1967. Experiments on density and turbidity currents, 3. Deposition of sediment. *Can. J. Earth Sci.*, 4: 475–505.

Minter, W. E. L., 1970. Origin of mud polygons that are concave downward. *J. Sediment. Petrol.*, 40: 755–756.

Moody-Stuart, M., 1966. High- and low-sinuosity stream deposits with examples from Devonian of Spitsbergen. *J. Sediment. Petrol.*, 36: 1101–1117.

Morisawa, M., 1968. *Streams.* McGraw-Hill, New York, N.Y., 175 pp.

Morris, R. C., 1971. Classification and interpretation of disturbed bedding types in Jackfork flysch rocks (Upper Mississippian) Ouachita Mountains, Arkansas. *J. Sediment. Petrol.*, 41: 410–424.

Mrakovich, J. V., 1969. *Fluvial Deposits of the Sharon Conglomerate in Portage, Summit, Eastern Medina, and Northeastern Wayne Counties.* Thesis, Kent State Univ., Kent, Ohio, 92 pp.

Newton, R. S., 1968. Internal structure of wave-formed ripple marks in the nearshore zone. *Sedimentology*, 11: 275–292.

Ore, H. T., 1964. Some criteria for recognition of braided steam deposits. *Contrib. Geol.*, 3: 1–14.

Osborne, F. F., 1953. Concretion conglomerate in the Charny Sandstone, Quebec. *Trans. R. Soc. Can.*, 47: 55–60.

Otto, G. H., 1938. The sedimentation unit and its use in field sampling. *J. Geol.*, 46: 569–582.

Otvos, E. G., Jr., 1965. Types of rhomboid beach surface patterns. *Am. J. Sci.*, 263: 271–276.

Peabody, F. E., 1947. Current crescents in the Triassic Moenkopi Formation. *J. Sediment. Petrol.*, 17: 73–76.

Pelletier, B. R., 1958. Pocono paleocurrents in Pennsylvania and Maryland. *Bull. Geol. Soc. Am.* 69: 1033–1064.

Pettijohn, F. J., 1957. *Sedimentary Rocks.* Harper, New York, N.Y., 718 pp.

Pettijohn, F. J. and Potter, P. E., 1964. *Atlas and Glossary of Primary Sedimentary Structures.* Springer, New York, N.Y., 370 pp.

Picard, M. D., 1966. Oriented, linear-shrinkage cracks in Green River Formation (Eocene), Raven Ridge area, Uinta Basin, Utah. *J. Sediment. Petrol.*, 36: 1050–1057.

Picard, M. D., 1967a. Paleocurrents and shoreline orientations in Green River Formation (Eocene), Raven Ridge and Red Wash areas, northeastern Uinta Basin, Utah. *Bull. Am. Assoc. Petrol. Geologists*, 51: 383–392.

Picard, M. D., 1967b. Paleocurrents and shoreline orientations in Green River Formation (Eocene), Raven Ridge and Red Wash areas, northeastern Uinta Basin, Utah: reply to discussion by R. A. Davis, Jr. *Bull. Am. Assoc. Petrol. Geologists*, 51: 2471–2475.

Picard, M. D., 1969. Oriented linear-shrinkage cracks in Alcova Limestone Member (Triassic), southeastern Wyoming. *Contrib. Geol.*, 8: 1–7.

Picard, M. D., 1970a. Relative abundance of bedding types and sedimentary structures formed by straight to slightly meandering ephemeral streams. *J. Sediment. Petrol.*, 40: 772 (abstr.).

Picard, M. D., 1970b. Current directions in straight to slightly meandering ephemeral streams. *J. Sediment. Petrol.*, 40: 771–772 (abstr.).

Picard, M. D. and High, L. R., Jr., 1964. Pseudo rib-and-furrow marks in the Chugwater (Triassic) Formation of west-central Wyoming. *Contrib. Geol.*, 3: 27–31.

Picard, M. D. and High, L. R., Jr., 1968. Shallow-marine currents on the Early (?) Triassic Wyoming shelf. *J. Sediment. Petrol.*, 38: 411–423.

Picard, M. D. and High, L. R., Jr., 1969. Some sedimentary structures resulting from flash floods. *Bull. Utah. Geol. Mineral. Surv.*, 82: 175–190.

Picard, M. D. and High, L. R., Jr., 1970a. Interference ripple marks formed by ephemeral streams. *J. Sediment. Petrol.*, 40: 708–711.

Picard, M. D. and High, L. R., Jr., 1970b. Sedimentology of oil-impregnated, lacustrine and fluvial sandstone, P.R. Spring area, southeast Uinta Basin, Utah. *Utah Geol. Mineral. Surv., Spec. Stud.*, 33: 32 pp.

Picard, M. D. and High, L. R., Jr., 1971. Sedimentary structures and bedding along ephemeral streams. *Bull. Am. Assoc. Petrol. Geologists*, 55: 357–358 (abstr.).

Picard, M. D. and Hulen, J. B., 1969. Parting lineation in siltstone. *Bull. Geol. Soc. Am.*, 80: 2631–2636.

Potter, P. E. and Pettijohn, F. J., 1963. *Paleocurrents and Basin Analysis.* Academic Press, New York, N.Y., 296 pp.

Power, W. R., Jr., 1961. Backset beds in the Coso Formation, Inyo County, California. *J. Sediment. Petrol.*, 31: 603–607.

Reiche, P., 1950. A survey of weathering processes and products. *Univ. New Mex. Publ. Geol.*, 3: 95 pp.

Reineck, H.-E., 1961. Sedimentbewegungen an Kleinrippeln im Watt. *Senckenbergiana Lethaea*, 42: 51–67.

Reineck, H.-E., 1963. Sedimentgefüge im Bereich der südlichen Nordsee. *Abh. Senckenberg. Naturforsch. Ges.*, 505: 1–136.

Reineck, H.-E., and Wunderlich, F., 1968. Zur Unterscheidung von asymmetrischen Oszillationrippeln und Strömungsrippeln. *Senckenbergiana Lethaea*, 47: 321–345.

Richter, R., 1926. Die Entstehung von Tongeröllen und Tongallen unter Wasser. *Senckenbergiana*, 8: 305–315.

Sanders, J. E., 1960. Origin of convoluted laminae. *Geol. Mag.*, 97: 409–421.

Schlee, J., 1957. Fluvial gravel fabric. *J. Sediment. Petrol.*, 27: 162–176.

Schumm, S. A., 1960a. The shape of alluvial channels in relation to sediment type. *U.S. Geol. Surv., Profess. Pap.*, 352-B: 17–30.

Schumm, S. A., 1960b. The effect of sediment type in the shape and stratification of some modern fluvial deposits. *Am. J. Sci.*, 258: 177–184.

Schumm, S. A., 1961. Effect of sediment characteristics in erosion and deposition in ephemeral-stream channels. *U.S. Geol. Surv., Profess. Pap.*, 352C: 31–70.

Schumm, S. A., 1963. Sinuosity of alluvial channels on the Great Plains. *Bull. Geol. Soc. Am.*, 74: 1089–1100.

Schumm, S. A., 1968. River adjustment to altered hydrologic regimen–Murrumbidgee River and paleochannels, Australia. *U.S. Geol. Surv., Profess. Pap.*, 598: 65 pp.

Seilacher, A., 1964. Biogenic sedimentary structures. In: *Approaches to Paleoecology.* Wiley, New York, N.Y., pp.296–326.

Selley, R. C., Shearman, D. J., Sutton, J. and Watson, J., 1963. Some underwater disturbances in the Torridonian of Skye and Raasay. *Geol. Mag.*, 100: 224–243.

Sharp, R. P., 1963. Wind ripples. *J. Geol.*, 71: 617–636.

Shrock, R. R., 1948. *Sequence in Layered Rocks.* McGraw-Hill, New York, N.Y., 507 pp.

Siever, R., 1951. The Mississippian Pennsylvanian unconformity in southern Illinois. *Bull. Am., Assoc. Petrol. Geologists*, 35: 542–581.

Simons, D. B. and Richardson, E. V., 1961. Forms of bed roughness in alluvial channels. *Proc. Am. Soc. Civil Eng., J. Hydraulics Div.*, 87: 87–105.

Simons, D. B. and Richardson, E. V., 1963. Forms of bed roughness in alluvial channels. *Trans. Am. Soc. Civil Eng.*, 128: 284–323.

Simons, D. B., Richardson, E. V. and Nordin, C. F., Jr., 1965. Sedimentary structures generated by flow in alluvial channels. *Soc. Econ. Paleontologists Mineralogists, Spec. Publ.*, 12: 34–52.

Smith, H. T. U., 1965. Wind-formed pebble ripples in Antarctica. *Geol. Soc. Am. Progr. Ann. Meeting*, p.157.

Smith, N. D., 1970. The braided stream depositional environment: comparison of the Platte River with some Silurian clastic rocks, north-central Appalachians. *Bull. Geol. Soc. Am.*, 81: 2993–3014.

Smith, N. D., 1971a. Pseudo-planar stratification produced by very low amplitude sand waves. *J. Sediment. Petrol.*, 41: 69–73.

Smith, N. D., 1971b. Transverse bars and braiding in the Lower Platte River, Nebraska. *Bull. Geol. Soc. Am.*, 82: 3407–3420.

Sorauf, J. E., 1965. Flow rolls of Upper Devonian rocks of south-central New York state. *J. Sediment. Petrol.*, 35: 553–563.

Sorby, H. C., 1908. On the application of quantitative methods to the study of the structure and history of rocks. *Q. J. Geol. Soc. Lond.*, 64: 171–232.

Spreng, A. C., 1967. Slump features, Fayetteville Formation, northwestern Arkansas. *J. Sediment. Petrol.*, 37: 804–817.

Stanley, D. J., 1964. Large mudstone–nucleus sandstone spheroids in submarine channel deposits. *J. Sediment. Petrol.*, 34: 672–676.

Stanley, D. J., 1968. Graded bedding-sole marking-graywacke assemblage and related structures in some Carboniferous flood deposits, eastern Massachusetts. In: *Late Paleozoic and Mesozoic Continental Sedimentation, Northeastern North America–Geol. Soc. Am., Spec. Pap.*, 106: 211–239.

Stanley, D. J., 1969. Armored mud balls in an intertidal environment, Minas Basin, southwest Canada. *J. Geol.* 77: 683–693.

Stewart, A. D., 1963. On certain slump structures in the Torridonian sandstones of Applecross. *Geol. Mag.*, 100: 205–218.

Stokes, W. L., 1947. Primary lineation in fluvial sandstones. *J. Geol.*, 55: 52–54.

Swartz, J. H., 1927. Subaerial sun-cracks. *Am. J. Sci.*, 14: 69–70.

Tanner, W. F., 1960. Shallow-water ripple mark varieties. *J. Sediment. Petrol.*, 30: 481–485.

Tanner, W. F., 1964. Eolian ripple marks in sandstone. *J. Sediment Petrol.*, 34: 432–433.

Tanner, W. F., 1966. Numerous eolian ripple marks from Entrada Formation. *Mountain Geol.*, 3: 133–134.

Tanner, W. F., 1967. Ripple-mark indices and their uses. *Sedimentology*, 9: 89–104.

Teichert, C., 1970. Runzelmarken (wrinkle marks). *J. Sediment. Petrol.*, 40: 1056–1057.

Termier, H. and Termier, G., 1963. *Erosion and Sedimentation.* Van Nostrand, London, 433 pp.

Trefethen, J. M. and Dow, R. L., 1960. Some features of modern beach sediments. *J. Sediment. Petrol.*, 30: 589–602.

Twenhofel, W. H., 1921. Impressions made by bubbles, rain-drops, and other agencies. *Bull. Geol. Soc. Am.*, 32: 359–372.

Twenhofel, W. H., 1932. *Treatise on Sedimentation.* Williams and Wilkins, Baltimore, Md., 2nd ed., 926 pp.

Van Houten, F. B., 1964. Cyclic lacustrine sedimentation, Upper Triassic Lockatong Formation, central New Jersey and adjacent areas. *Kansas Geol. Surv., Bull.*, 169: 497–531.

Van Straaten, L. M. J. U., 1951. Longitudinal ripple marks in mud and sand. *J. Sediment. Petrol.*, 21: 47–54.

Van Straaten, L. M. J. U., 1954. Sedimentology of recent tidal-flat deposits and the Psammites du Condroz (Devonian). *Geol. Mijnbouw*, 16: 25–47.

Visher, G. S., 1965. Use of vertical profile in environmental reconstruction. *Bull. Am. Assoc. Petrol. Geologists*, 49: 41–61.

Visher, G. S., 1969. Grain size distributions and depositional processes. *J. Sediment. Petrol.*, 39: 1074–1106.

Waechter, N. B., 1970. Braided stream deposits of the Red River, Texas Panhandle. *Geol. Soc. Am., Progr. Ann. Meeting*, 2: 713.

Walker, R. G., 1963. Distinctive types of ripple-drift cross-lamination. *Sedimentology*, 2: 173–188.

Walker, R. G., 1969. Geometrical analysis of ripple-drift cross-lamination. *Can. J. Earth Sci.*, 6: 383–391.

Ward, F., 1923. Note on mud cracks. *Am. J. Sci.*, 6: 308–309.

Wheeler, H. E. and Quinlan, J. L., 1951. Precambrian sinuous mud cracks from Idaho and Montana. *J. Sediment. Petrol.*, 21: 141–146.

White, W. A., 1961. Colloid phenomena in sedimentation of argillaceous rocks. *J. Sediment. Petrol.*, 31: 560–570.

White, W. S., 1952. Imbrication and initial dip in a Keweenawan conglomerate bed. *J. Sediment. Petrol.*, 22: 189–199.

Williams, G. E., 1966. Planar cross-stratification formed by the lateral migration of shallow streams. *J. Sediment. Petrol.*, 36: 742–746.

Williams, G. E., 1968. Formation of large-scale trough cross-stratification in a fluvial environment. *J. Sediment. Petrol.* 38: 136–140.

Williams, G. E., 1971. Flood deposits of the sand-bed ephemeral streams of central Australia. *Sedimentology*, 17: 1–40.

Williams, P. F. and Rust, B. R., 1969. The sedimentology of a braided river. *J. Sediment. Petrol.*, 39: 649–679.

Wolf, K. H., 1965. Petrogenesis and paleoenvironments of Devonian algal limestones of New South Wales. *Sedimentology*, 4: 113–178.

Woodford, A. O., 1935. Rhomboid ripple mark. *Am. J. Sci.*, 29: 518–525.

Wunderlich, F., 1970. Genesis and environment of the "Nellenköpfchenschichten" (Lower Emsian, Rheinian Devon) at locus typicus in comparison with modern coastal environment of the German Bay. *J. Sediment Petrol.*, 40: 102–130.

Yeakel, L. S., Jr., 1962. Tuscarora, Juniata, and Bald Eagle paleocurrents and paleogeography in the central Appalachians. *Bull. Geol. Soc. Am.*, 73: 1515–1539.